So Great a Vision

Middlebury Bicentennial Series in Environmental Studies

Christopher McGrory Klyza and Stephen C. Trombulak, *The Story of Vermont: A Natural and Cultural History*

Elizabeth H. Thompson and Eric R. Sorenson, *Wetland, Woodland, Wildland: A Guide to the Natural Communities of Vermont*

John Elder, editor, *The Return of the Wolf: Reflections on the Future of Wolves in the Northeast*

Kevin Dann, *Lewis Creek Lost and Found*

Christopher McGrory Klyza, editor, *Wilderness Comes Home: Rewilding the Northeast*

Terry Osborne, *Sightlines: The View of a Valley through the Voice of Depression*

Stephen C. Trombulak, editor, *So Great a Vision: The Conservation Writings of George Perkins Marsh*

So Great a Vision

The Conservation Writings of George Perkins Marsh

Edited by

Stephen C. Trombulak

Middlebury College Press
Published by University Press of New England
Hanover and London

Middlebury College Press
Published by University Press of New England, Hanover, NH 03755
© 2001 by The President and Trustees of Middlebury College
All rights reserved
Printed in the United States of America

5 4 3 2 1

Contents

Foreword by John Elder   vii

Introduction   ix

Address Delivered before the Agricultural Society of   5
Rutland County (September 30, 1847)

Lectures Delivered before the Smithsonian Institution,   24
No. 1—The Camel (1855)

Oration before the New Hampshire State Agricultural   34
Society (October 10, 1856)

Report, Made under Authority of the Legislature of   62
Vermont, on the Artificial Propagation of Fish (1857)

The Study of Nature (1860)   73

Irrigation: Its Evils, Remedies, and the Compensations (1874)   98

MAN AND NATURE; OR, PHYSICAL GEOGRAPHY AS
MODIFIED BY HUMAN ACTION (1864)

Chapter I: Introductory   123

Chapter II: Transfer, Modification, and Extirpation of   140
Vegetable and of Animal Species

Chapter III: The Woods   159

Chapter IV: The Waters   187

Chapter V: The Sands   207

Chapter VI: Projected or Possible Geographical Changes by Man   217

Suggested Readings   227

Foreword

Man and Nature has been a landmark in environmental thought ever since George Perkins Marsh published it in 1864. By demonstrating the importance of forests to ecological and social health, Marsh contributed directly to the founding of both the Adirondack Park and the national park system. His writings were also often referred to in the debate leading up to the Wilderness Act of 1964. Yet the monumental scope of his masterpiece and the unusual density of its prose have sometimes made Marsh's book more talked about than actually read. Even in college courses on environmental history and philosophy, it often makes its way onto syllabi only through brief excerpts. Today, though, when conservationists strive to achieve a more ecologically and socially inclusive perspective, the comprehensiveness of Marsh's writing can again be enormously helpful.

One central purpose of *So Great a Vision: The Conservation Writings of George Perkins Marsh* is to offer selections from *Man and Nature* that are at once substantial, representative, and manageable within most college courses. This collection also attempts to locate Marsh's masterpiece in the context of his broader intellectual development. The editor, Stephen Trombulak, has accordingly chosen to lead up to his selections from the book with half a dozen of the author's most important essays and addresses. In addition to their inherent historical interest, these pieces illuminate central concerns of Marsh's that may occasionally be obscured by his book's sheer magnitude and by its wealth of documentation. They orient a reader to the perceptiveness of his observations, as well as to the power and originality of his argument.

Marsh grew up in Woodstock, Vermont, at the beginning of the nineteenth century, an era when deforestation was rampant and such negative consequences as erosion and impaired waterways were all too evident. In his career as an American diplomat to Turkey and Italy, he was able to connect his childhood landscape with the much vaster history of settlement and agriculture around the Mediterranean. Stimulated by such cross-cultural experiences, Marsh was able to synthesize many principles of forest health, sustainable agriculture, and hydrology that continue to be valid and illuminating today.

Marsh's richly stocked mind anticipated what we sometimes refer to today as a bioregional perspective. Such an approach focuses on the integrity of watersheds and other naturally delineated domains, and it sees human culture as both reflecting and influencing physical landscapes. Well over a century since its original publication, *Man and Nature* remains both a uniquely comprehensive and a remarkably bold example of such thinking. If we can return now, not only to what Marsh meant to American conservation, but also to exactly what he *said*, we may find in him an inspiring ancestor and an encouraging guide.

John Elder
Middlebury College

Introduction

The fabric of social discourse today is strongly influenced by our concerns over a host of major environmental issues. It is impossible to read a newspaper or pass through an election year without being confronted with a frightening array of questions about how our societies, both national and international, should best solve problems ranging from air and water pollution to the loss of tropical rainforests, from global warming to the thinning of the ozone layer, and from the depletion of freshwater supplies to soil erosion. From our modern vantage point, it is easy to forget that all debates about environmental policies have as a fundamental tenet an acceptance that it *is* actually possible for humans to transform their environment in ways that reduce the quality of their lives. The widespread realization that we can alter our environment in negative ways is one of the most significant intellectual transformations in how we view our relationship with nature.

No one person can take all the credit for effecting this transformation. Throughout the nineteenth and twentieth centuries, many people argued that the status quo environmental philosophy—a philosophy that assumed our numerical growth and technological development was, at worst, benign—was simply wrong.

But George Perkins Marsh, the nineteenth-century American scholar, author, and statesman, with the publication in 1864 of his book *Man and Nature; or, Physical Geography as Modified by Human Action*, did more than any other single person to make clear to a broad audience society's widespread environmental impacts and our need to guide social development in light of this knowledge. The historian Lewis Mumford called *Man and Nature* "the fountain-head of the conservation movement."

George Perkins Marsh was the quintessential Renaissance man. During the course of his long and productive life he was, at various stages and with different degrees of success, a lawyer, businessman, inventor, scholar of languages (coming to speak at least twenty), art collector, politician, bureaucrat, diplomat, explorer, and architect. But his enduring legacy is based on his writings about the relationship between humans and their environment, writings that have at their core observations and perspectives that he

A sketch by H. P. Moore (1859) of Woodstock, Vermont, viewed from Mount Tom—showing the landscape as Marsh would have seen it as a boy.

gained as a boy growing up in eastern Vermont, supplemented by a lifetime of scholarly reading and international travel.

Marsh was born in 1801 in Woodstock, Vermont. Woodstock, situated in the Ottauquechee River valley close to its confluence with the Connecticut River, was among the first heavily settled regions in Vermont. By the early 1800s much of the forestland in the area had been cleared for hill and valley-bottom farms. Even as a young boy, Marsh was permanently impressed by the impact that forest clearing was having on hillside erosion, frequency of wildfires, and control of surface-water flow. He saw firsthand the enormous importance of standing forests to a host of related ecological conditions, and hence the role of human action, since humans were the primary cause of large-scale deforestation. As a boy, he repeated his observations of the effects of forest clearing numerous times as he traveled widely with his father, Charles, a prominent lawyer and one of Woodstock's most notable citizens. His father helped encourage Marsh's interest in the natural world, teaching him the names of trees, the concept of watersheds, and a range of other topics in geography and natural history.

Yet as much as Marsh loved the out-of-doors, his most enduring character trait was his love of books and reading. He had a bright and eager mind, leading him to graduate from Dartmouth College in nearby Hanover, New Hampshire, when he was only nineteen. By that time he was fluent in Greek, Latin, Portuguese, Spanish, French, and Italian, being largely self-taught in them all. After a brief stint as an English and language teacher at the Norwich Academy in Vermont, he returned to Woodstock, where he studied law with his father. Marsh soon passed the bar exam and moved to Burlington, Vermont, where he entered into a law partnership with the energetic and outgoing B. F. Bailey in 1825. Burlington then, as now, was a center for trade and industry, and their small firm enjoyed great success. Yet with Bailey's untimely death in 1832, Marsh's weaknesses as a businessman and his general dislike for the people he had to associate with as a lawyer led to a steady decline in the business until he finally closed the office in 1842.

In 1828, Marsh married Harriet Buell, who sadly died of a heart ailment only five years later. They had two sons, the oldest of which died of scarlet fever only a few days after his mother's death. The younger boy, George Ozias, was sent to Woodstock to live with Marsh's mother. Although he would live with Marsh on and off through the succeeding years, he died in his early thirties after years of illness, alcoholism, and bitter relations with his father. Marsh was despondent after Harriet's death, a state that contributed to the decline of his law practice, and for many years he isolated himself from society and spent his time in the study of Scandinavian languages. Eventually, however, Marsh emerged from his emotional exile and began to mix again with the circle of learned men at the University of Vermont in Burlington. One of his friends at that time was Zadoch Thompson, the author of the first authoritative natural historical account of Vermont, who helped rekindle Marsh's appreciation for the natural world.

Marsh eventually fell in love again. He remarried in 1839 to Caroline Crane, who despite her own poor health was his near-constant companion and source of emotional support for the remaining forty-three years of his life. At about this time, Marsh's venture into wool manufacturing ended disastrously, the first of his many failed business dealings that would eventually result in bankruptcy and then near-constant worries about money and anticipated poverty for the rest of his life.

In part to escape to a better climate and in part because he had few other employment opportunities, Marsh ran for and was elected to one of Vermont's four congressional seats in the U.S. House of Representatives in 1843, where he served until 1849. His political philosophies were formed while he served a short term in the Vermont Legislature in the mid-1830s. He viewed himself as democratic and, although he had a poor opinion of the common person, egalitarian. He also saw himself as practical, believing passionately in progress, utility, and technology. As a member of the Whig party in Congress, he campaigned stridently against both slavery and the admission of Texas into the Union, and for high tariffs. At that time, however, the Whigs were the minority party in Congress, and Marsh's views

Portrait of George Perkins Marsh painted by G. P. A. Healy (1844).

were in strict opposition to those of the majority Democrats; so he was able to accomplish very little. Perhaps his greatest achievement came from the influential role he played in helping to establish the Smithsonian Institution, which remains today as perhaps the greatest research and museum center in the world.

From the perspective of Marsh's own life, however, his years in Washington, D.C., had their greatest impact through the professional contacts he developed, contacts that resulted in his spending most of the next thirty-three years as a foreign diplomat for the U.S. government, first as the U.S. Minister to Turkey and then, after seven years back in Vermont, as the Minister to the new nation of Italy.

For several months during his first year as Minister to Turkey, Marsh and his wife traveled widely through Egypt and Arabia, a journey that formed the basis of Marsh's observations about the magnitude and type of human modifications of desert environments, a theme he developed at great length more than a decade later in his book *Man and Nature*. It was during this trip that he also observed the widespread use of camels as domestic and military animals, a development that he later argued the United States should adopt for its own military uses in its arid West. His role as Minister subsequently took him to Greece, which gave him the easy opportunity to explore the Alps, an area he was to return to, study, and write about for many years after, even up to the time of his death.

By 1853 he had been replaced in Turkey and had returned to the United States, eventually to settle once again in Vermont as the state's Railroad Commissioner, Statehouse Commissioner, and Fish Commissioner. It was during this time that he actively promoted government control of certain social institutions, most notably transportation and communication, a view developed in large response to his recent financial losses at the hands of unscrupulous private railroad interests. As Fish Commissioner, he also authored an influential and visionary report for the Vermont Legislature on the artificial propagation of fish.

It was also during these years in Vermont that the outline of the themes he would develop in *Man and Nature*, his greatest and most lasting work, and one of the most influential environmental books in history, began to take form in his mind. The observations he had made as a boy in Vermont and more recently in Arabia and Europe led him to develop a new perspective on human geography. Marsh noted that whether he looked at forests, mountains, waters, or deserts, he saw everywhere irrefutable evidence that humanity had a profound effect on the physical environment of the earth. His extensive reading provided him with abundant examples, drawn from fields ranging from botany to geology to archeology to civil engineering, that supported his central thesis. Man (to use Marsh's language) was not simply a passive inhabitant of the natural world. He was a potent force of change, for both good and ill.

In 1861, Marsh once again gained a diplomatic post, being appointed by President Lincoln as the U.S. Minister to Italy, a post retained until his

Marsh with his wife Caroline, right, and Caroline's sister, Lucy Crane, center (ca. 1840).

death twenty-one years later. Marsh completed work on the first edition of *Man and Nature* in 1864 while living near Turin. For the rest of his life, Marsh traveled widely throughout Italy, revised *Man and Nature* twice, wrote a report for the U.S. Commission of Agriculture on the social and environmental consequences of irrigation, particularly in the newly opened

lands of the American arid West, and developed a fascination for forestry, the subject he was pursuing at the time of his death.

George Perkins Marsh was a complex man, at various times in his life holding views that today would be regarded as jingoistic, racist, mildly sexist, and religiously chauvinistic. Yet time and experience led Marsh to blunt and recant many of his earlier intolerances. He was also capable of espousing dramatically contradictory views. For example, he could, on the one hand, recount the problems associated with the introduction of species outside their native ranges as well as, on the other hand, promote the introduction of the camel into the United States and a host of American plants and animals into Italy. Marsh can perhaps be best understood as a man who believed quite strongly in the capabilities of ordinary people, be they Vermont hill farmers, Egyptian nomads, or Italian vintners. His conservation writings can be seen as testimonies to his belief that human society could progress to greatness, but only if it paid attention to nature and learned the lessons it taught. His works, most notably *Man and Nature*, forever changed our culture's perception of our relationship with the natural world. Society's fundamental assumption could no longer be that the natural world simply shaped humanity. Based on his vision, the new understanding became that humanity could also shape the natural world. Our guiding question from that point on became, "What must we do to live within the boundaries established by nature's harmonies?"

Given his enormous impact on the shape of our environmental consciousness, it is surprising that so few people today have ever heard of George Perkins Marsh, and that so few even among those who have a deep understanding of environmental issues are aware of any of the details of Marsh's thinking. These deficiencies are, perhaps, best explained by aspects of Marsh's own writings. First, Marsh did not often publish his thoughts about the environment (or what he called "human geography"), and what he did publish was generally only in outlets that today are not widely available. The notable exception, of course, is *Man and Nature*, which remains in print in its entirety more than 130 years after it was first published. Yet his other environmental works, written from 1847 to 1874, have to my knowledge never been reprinted and remain largely unread.

Second, Marsh wrote in a style common to men and women of letters in the nineteenth century, which more modern readers often find dense, even impenetrable. Also, because he was developing a new environmental perspective and marshaling in its support a diverse range of facts and observations, he provided numerous examples for each point, leading to much repetition. As a result, *Man and Nature* remains one of the most influential but least read books written in the last two hundred years.

Marsh in his library in Florence, Italy (ca. late 1860s).

This volume seeks to correct these problems and to make Marsh's conservation writings more accessible to the present generation. I have included here all of Marsh's publications that reflect on his environmental thinking. With the exception of *Man and Nature*, I present them either in their entirety or edited to highlight just their portions of environmental relevance. I have, on the other hand, edited *Man and Nature* heavily, including here less than one-fifth of its original text. I edited out almost all of Marsh's extensive footnotes, as well as many of the examples he provided for his key points. Readers interested in Marsh's references or the complete argument developed in *Man and Nature* should read the original text, most recently reprinted by Harvard University Press and skillfully annotated by David Lowenthal, Marsh's primary biographer. I include a small number of Marsh's footnotes that contribute to the development of his theme. I have placed each in the main body of the text as a parenthetical comment at the location set by Marsh. In a few cases I have inserted into the text in brackets and italics my own comments, either for clarification or for correction. In all the selections I retained Marsh's original language but corrected errors that I felt were introduced by the printers or were noted as "manuscript corrections by the author" in H. L. Koopman's 1892 bibliography of George Perkins Marsh. I preface each selection with my own synopsis of its major

Portrait of George Perkins Marsh by J. E. Bartlett (ca. 1880).

themes, especially as they relate to modern conservation thinking. I believe that a comparison of Marsh's vision with today's reality amplifies our appreciation of the magnitude of his insights and influence.

This book came to life with the help of many people, most notably John Elder and Chris McGrory Klyza, my co-editors of the Middlebury College

Bicentennial Series in Environmental Studies, of which this book is a part. Even separate from our work together on this series, however, I am indebted to both of them for years of intellectual debate, which helped shape my ability to understand and appreciate Marsh's vision. Phil Pochoda and April Ossmann at the University Press of New England provided constant and invaluable editing and production advice. Susan Tucker, research librarian at Middlebury College, helped me track down many of Marsh's less readily available writings, including several that are not included in this volume. Joanna Shipley and Anne Every painstakingly transcribed Marsh's original works. My daughter, Sage Trombulak, helped me to proof this text against the originals. Much of my editing and writing for this book took place while I was on sabbatical leave at the University of Adelaide in Australia. I am indebted to Hugh Possingham and his cadre of postdoctoral fellows, graduate students, and honors students, particularly Ian Ball, Steve Ball, Clare Bradley, Scott Field, Drew Tyre, and Margy Wright, not only for creating a great intellectual environment but also for tolerating my frequent outbursts as I struggled through the complexities of Marsh's writing.

All photographs are reproduced courtesy of the Special Collections of the University of Vermont.

So Great a Vision

Address Delivered before the Agricultural Society of Rutland County

[Delivered September 30, 1847; published at the request of the Society
by the Rutland Herald, 1848]

Marsh's address to the Agricultural Society of Rutland County, Vermont, given while he was a U.S. Congressman, was his first speech that was later made available in print. Several themes that are constant throughout Marsh's later writings emerge in this brief presentation. The first of these is his belief in the superiority of humans over the rest of nature by divine right, and the second is his belief in the particular superiority of the cultures that derived from western Europe. Embedded within these anthropocentric and Eurocentric views, however, are also the kernels of ideas that presage the development of his magnum opus, *Man and Nature,* published seventeen years later, and also of many philosophical beliefs present in the environmental ethics of the late twentieth century.

This address, on the whole, was meant to be a paean to farmers, who during the early and mid-1800s were so central to the cultural and economic identity of Vermont. Unlike his contemporary Henry David Thoreau, Marsh bore no strong attachment to wild nature, referring to agriculture in this speech and elsewhere as a force that brings light and life to forests, and to landscapes without humans as unproductive.

Yet he also argues here in support of many environmental themes that are widely espoused today, including the importance of forests in the control of climate, soil nutrients, erosion, and exotic weeds; the impact of increasing human population size on the demand for food production; the importance of plants from tropical climates for U.S. and European agriculture (foreshadowing contemporary arguments linking the conservation of tropical forests and the agricultural economies of developed countries); the responsibility of humans not to abuse nature; the impact of human modification of the landscape on climate; the importance of conservation in the present for the well-being of future generations; the need for people to decrease their level of meat consumption and increase their intake of fruits and vegetables; and the importance of caring for our homes to foster an intergenerational sense of place. From the first, Marsh's environmental views were strictly utilitarian, placing everything in the context of its impact on humans, but they were passionately held and powerfully argued.

Although the Association, which I have the honor to address, is styled an Agricultural Society, its influence is not designed to be limited to the encouragement and improvement of the culture of the soil, but its objects are threefold, and embrace as well the toils of the herdsman and the mechanic as the labors of the ploughman. I shall, therefore, not be expected to confine my remarks within a narrower range than your sphere of operations, and while I shall make no attempt to lay down minute practical rules for the conduct or economy of either of these great branches of productive industry, I shall endeavor briefly to illustrate the importance of them all, considered as means and instruments of civilization and social progress, and shall suggest, in a general way, some improvements, the promotion of which seems to me an object well worthy the zealous efforts of the agricultural associations of Vermont.

Before I proceed to the discussion of this my proper subject, it may not be amiss to notice certain particulars connected with the early history, physical condition, and fundamental legislation, of the American Continent, and especially of the United States, which have had an important bearing on the prosperity of the industrial arts, and the social condition of those who have made them their vocation.

America offers the first example of the struggle between civilized man and barbarous uncultivated nature. In all other primitive history, the hero of the scene is a savage, the theatre a wilderness, and the earth has been subdued in the same proportion, and by the same slow process, that man has been civilized. In North America, on the contrary, the full energies of advanced European civilization, stimulated by its artificial wants and guided by its accumulated intelligence, were brought to bear at once on a desert continent, and it has been but the work of a day to win empires from the wilderness, and to establish relations of government and commerce between points as distant as the rising and the setting sun. This marvellous change, which has converted unproductive wastes into fertile fields, and filled with light and life the dark and silent recesses of our aboriginal forests and mountains, has been accomplished through the instrumentality of those arts, whose triumphs you are this day met to celebrate, and your country is the field, where the stimulus of necessity has spurred them on to their most glorious achievements. But besides the new life and vigor infused into these arts, by the necessity of creating food and shelter and clothing for a swarming emigration and a rapidly multiplying progeny, the peculiar character of the soil and of the indigenous products of America has introduced most important modifications into the objects and processes of all of them, by offering to European industry new plants for cultivation, and new and more abundant materials for artificial elaboration. At the same time, American husbandry and mechanical art are totally different in their objects, character, and processes from what they would be, were they conversant only with the indigenous products of our native soil. To exemplify; America has given to the Eastern Hemisphere maize, tobacco, the potato, the batata, the pine-apple, the turkey, and more lately the alpaca, not to mention innumerable flowering plants, as well as other vegetables of less economical importance, or the tribute which her peltry, her forests, her fisheries, and her mines of gems and the precious metals have paid to European cupidity; she has received in return wheat, rye, other cerealia, new varieties of the cotton plant, flax, hemp, rice, the sugar-cane, coffee, our orchard fruits, kitchen and medicinal roots, pulse and herbs, the silkworm, the honey-bee, the swine, the goat, the sheep, the horse and the ox. By these interchanges, the industry of both continents has been modified and assimilated, and it is a curious fact, that the greater proportion of properly agricultural American labor is devoted to the growth of vegetable products of transatlantic origin, while the workshops and the maritime commerce of Europe find one of their principal sources of employment in the conversion or the carriage of vegetable substances either indigenous to, or most advantageously grown in the soil of America. The colonization of a new continent under such remarkable circumstances could not fail to give a powerful impulse to the productive arts; and their increased economical, commercial, and financial importance has invested them with an interest in the eyes of statesmen, and a prominence as objects especially to be cherished, in every well regulated scheme of political economy, which they had never before attained, and the social position of those who are engaged in them has been elevated accordingly. Further, there are certain features of our institutions and our primary legislation, which have contributed not a little to raise and improve the condition of those who pursue the industrial arts, and especially of those devoted to agricultural occupations. The most important of these is the rejection of feudal tenures to lands, and the creation of pure allodial estates—a title scarcely known to the common law, and which makes every man the absolute irresponsible owner of his own land, subject neither to services, wardships, rents, tithes, reliefs, forfeiture, nor any manner of burden or restraint upon alienation by sale, inheritance or devise, or upon the cultivation of his soil for such purposes, or by such course of husbandry, as he may deem expedient. Another of these new features is the abolition of the law of primogeniture, and the equal distribution of all the estate of intestates, whether real or personal, between the representatives in equal degree, without distinction of age or sex. The effect of this system, together with the low price of lands, has been to make almost every person, who lives to years of maturity, an absolute proprietor of the soil, or in other words one of the landed nobility of the republic, for the notion of hereditary nobility in Europe was founded on the right of inheriting real estate, they who owned the soil of a particular country being considered as its rightful lords and governors, because, by concert among themselves, they could lawfully exclude all others from the right of possession, or ever of commorancy, upon any portion of its territory. Our laws do not indeed restrict political franchises to those alone, who are seized of real estate, but as a majority of those who are of the legal age for the exercise of those franchises are landholders, the proprietors of the soil are in fact here, as in most civilized governments, the real rulers of the land. The mechanic arts too, have been relieved from the burden of long apprenticeships, and other legal obstacles to their free exercise, and every species of productive industry is among us as free and unrestricted as the winds of heaven. The result of all this has been, that the arts of production as well as of conversion, in our time, and especially in our land, have proved a source of thrift to those who pursue them, of physical and financial strength to the commonwealth, and of general benefit to society, in a degree of which history gives no previous example, and they need only a wise, liberal and stable policy on the part of our government, to be a most important agent, in elevating us to as high a pitch of power and prosperity, as has as yet been attained by any nation under heaven.

I will now proceed to compare and illustrate in brief detail the relative value and importance of the three great divisions of productive labour, as means and instruments of civilization and social progress, first however glancing at the characteristic economic distinction between savage and civilized life.

In purely savage life, the wants of man are supplied by the destruction of the fruit, or plant, or animal, which clothes or feeds the human beast of prey, and while stripping the textile filaments from their vegetable stalk, or flaying and devouring the game which has fallen into his snare, he takes no thought for the reproduction of that which he improvidently consumes, but trusts implicitly to the bounty of spontaneous nature to supply the demands which the appetites and needs of her own children have created. Civilization begins with arrangements for securing the continued and regular supply of man's two great physical wants, food and clothing, by natural reproduction aided and promoted by artificial contrivances; and the degree of perfection to which these arrangements are carried, if it does not constitute the essence, at least furnishes a safe and convenient measure of the pitch of civilization, which a given people has attained. The arts of the savage are the arts of destruction; he desolates the region he inhabits, his life is a warfare of extermination, a series of hostilities against nature or his fellow man, and his labors are confined to the fabrication of weapons for slaying or repelling other tribes that intrude upon his hunting grounds, or of engines for ensnaring or destroying the wild animals on which he feeds. Civilization, on the contrary, is at once the mother and the fruit of peace. Social man repays to the earth all that he reaps from her bosom, and her fruitfulness increases with the numbers of civilized beings who draw their nutriment and clothing from the stores of her abundant harvests. The fowls of the air, too, and the beasts of the field, find in the husbandman a cherishing friend. The forest depths remote from the haunts of men yield sustenance to but a few of the tribes of animated nature. They are traversed only by swift-footed beasts, or strong winged birds of prey, and the humbler quadrupeds and gentler birds follow the migrations of the colonists, and gather upon the borders of civilization, where the abundance and variety of vegetable life affords them food, and the fear of man secures them protection against the ravages of the more rapacious brutes. Savage man then is the universal foe, both of his own kind and of all inferior organized existences, and incarnation of the evil principle of productive nature; civilization transforms him into a beneficent, a fructifying, and a protective influence, and makes him the monarch not the tyrant of the organic creation.

The objects on which industrial art is exercised in its first dawnings vary according to climate and the natural productions. In most countries of the old world, the germ of civilization with its attendant arts, is to be found in the pastoral life; in America, on the contrary, where nature with some doubtful and insignificant exceptions, denied to man the use of beasts of draught and burden, it began with agriculture, which in the Eastern Hemisphere constituted the second step of social advancement, while in some few favored insular climes, where the spontaneous productions of the soil and the sea suffice for human nourishment, the mechanic arts, which elsewhere form the last and crowning stage of material progress, have indicated the first movement of transition from savage to civilized existence. The enterprise of modern travellers has pushed discovery to the very verge of terrestrial creation, and colonization has followed so closely upon the heels of the voyager, that the arts of Europe have pervaded almost the whole habitable globe, and there are now extant but scanty remains of strictly savage life. The few tribes that have imbibed no tincture of civilization chiefly inhabit regions too frigid or too sterile for cultivation, and which deny both food and shelter to the domestic animals of Europe, while the pastoral races exist only as classes, in countries depending mainly on other branches of industry, or are confined to barren wastes like the sands of Arabia, the bleak mountains of Lapland, or the scarcely less desert steppes of Eastern Europe and Asia. Pastoral life admits of but a low state of civilization. The shepherd or herdsman is of necessity migratory in his habits, and must follow his flock wherever pasturage or water are most abundant; his occupancy of the soil, though temporary, must be exclusive; supplied by his herds with food and clothing and tent-cloth; he is unconscious of dependence upon his fellow man, and his social relations can be neither numerous nor close; he can recognize no superior but the chief of his sept, or rather his family, and all government must be simply patriarchal influence. The industrial processes belonging to the pastoral condition are few and simple, admitting of little variety, and affording narrow room for improvement. The habits of nomad tribes, therefore, lack the great element of civilization, progress; and the Bedouin of the Desert is unchanged since the days of Autar or of Job.

In civilized countries, as I have already intimated, pastoral life exists only in conjunction with agriculture, and these two branches of husbandry are in a great degree dependant upon each other. The improvement in the mechanic arts have not enabled the farmer to dispense with beasts of draught or burden. The steam plough is still but a dream, and the utility of even the mowing machine has not yet been established by successful experiment. The locomotive engine has not thus far been adapted to common roads, and notwithstanding the brilliant anticipations of some projectors, the day has not yet dawned, when every producer shall be whirled to market by the steam of his own teakettle. Nor has the increased production of vegetable aliment and tissues in any degree lessened the demand for animal food, or superseded the necessity of employing the skin and fur and wool of animals for clothing. On the other hand, the growing and gathering of the winter supply of food for the numerous flock and herds, which are required for our nourishment and clothing, constitutes a larger proportion of the labors of the agriculturist, and thus neither of these fields of rural husbandry can be much enlarged without a corresponding extension of the other. Pastoral life and the tillage of the earth are therefore no longer distinct and independent occupations, but are properly branches of the same calling, and though with some violence to etymology, popularly are, and conveniently may be, both comprehended under the general term AGRICULTURE.

Pure pastoral life, as I have said, advances man to but an humble stage of civilization, but when it is merged in agriculture, and the regular tillage of the soil commences, he is brought under the dominion of new influences, and the whole economy of domestic and social life is completely revolutionized.—Proper and permanent social institutions now begin to germinate, because combined effort for supplying the physical wants of man becomes necessary, and it is upon the necessity of such effort, that human society, considered as an artificial system, is founded. Men now begin to realize what, as wandering shepherds, they had before dimly suspected, that man has a right to the use, not the abuse, of the products of nature; that consumption should everywhere compensate, by increased production; and that it is a false economy to encroach upon a capital, the interest of which is sufficient for our lawful uses.

Among the various causes, by which the transition from the pastoral to the agricultural state may be occasioned or facilitated, an obvious one is the discovery, that by cultivation a smaller extent of ground may be made to furnish nourishment for the shepherd and his flocks, whose increasing numbers threaten to exhaust the supply spontaneously produced within his habitual range. Cultivation at once begun, the nomad condition is soon at an end, for the growing of a crop, including the preparation of the ground and the securing of the harvest, consumes the greater part of a year, and if perennial plants are reared, or the ground is first to be cleared of a forest growth, and fenced against the ravages of wild or domestic animals, many seasons must be passed upon the same spot. Hence arises the necessity of fixed habitations and store houses, and of laws which shall recognize and protect private exclusive right to determinate portions of the common earth, and sanction and regulate the right of inheritance, and the power of alienation and devise, in short the whole frame work of civil society. The recognition of private rights to real estate is a necessary condition precedent to the establishment of fixed habitations, without which there can be no permanent improvement of the soil, no considerable accumulation of personal property or the comforts of life, none of the sacred influences of home, no attachment to localities, no national feeling, no durable records, no history; it lies at the basis of all civil institutions; it is essential to the very idea of a nation, as distinguished from a nomadic tribe or sept; and those speculators, who propose the abrogation of such rights, are aiming a blow, not at the arbitrary institutions of an age or an epoch, but at fundamental principles on which society itself is grounded.

What then are the present condition and future prospects of the profession of Agriculture; how far it is to be regarded as a liberal art, and what part is it destined to play hereafter in the organization of the social fabric?

The primary and immediate object of agriculture is the cultivation of vegetables, and in its largest sense it embraces the care of the forest and the propagation of timber-trees, the rearing of fruits, of the roots and herbs and seeds of our gardens, and of our medicinal plants, as well as the growing of our ordinary farm crops. It may be stated as a general rule, subject to no clearly ascertained exceptions, that animated creatures are incapable of deriving any nutriment directly from mineral or unorganized matter. It is the great office of vegetable life to convert the ultimate elements of inorganic matter into secondary forms, sometimes called *proximate* or *immediate principles,* which either immediately, or as constituents or more complicated combinations (as roots, barks, leaves, fruits and seeds,) supply food, shelter, and clothing to man and the lower animals. In many instances again this process is repeated, the brute or insect being the agent of a second elaboration, and converting, by organic chemistry, vegetables unnutritious to higher creatures into animal fibre or other tissue capable of yielding them aliment, or subserving their other uses. And herein careful Nature has by no means left improvident man dependant upon the fruits of his own foresight and industry; for not only does he, reaping where he has not sown, derive supplies for his various wants from spontaneous vegetation, and from wild animals fed by that wild growth, but even in those uncounted ages of vital existence, which geologists tell us preceded his birth-day, she was busy in preparing and furnishing the future home of her noblest offspring. The preadamite world was clothed with a luxuriant growth of vegetable life, which fed and sheltered a swarming host of beast and bird and fish and creeping thing. But these were all tribes, unsuited to human use, and among the remains of primeval life, which lie sepulchred in the earth's crust, we find no plant or animal belonging to genera that yield food or clothing to man, no pleasant fruit, or nutritious seed or healthful root, no fowl or quadruped allied to our domestic animals, but rank ferns, unfruitful shrubs and barren evergreens, fishes of coarse and bony structure, huge, thick-skinned quadrupeds, and monstrous reptiles. And when, in the fullness of time appointed by the Creator, man was about to be summoned to assume dominion over the earth and all things upon it, the existing organic forms were swept away by the agency of some sudden catastrophe, or some unknown cause of gradual extinction, and succeeded by a more fruitful world of vegetable and animal life adapted to the convenience of him who was now called to reign over it. But though these extinct organizations were not, in their original forms, suited to human uses, they are yet, in their mineralized condition, of the highest value, and even of almost indispensable importance to man. The vast deposits of coal in Europe and America, without which the smelting of ores and the workings of metals could be practiced only on a very limited scale, are the remains of extinct forests; limestone formations are often almost entirely of animal origin, and even our common polishing powders are composed of the flinty shells or wing-cases of microscopic insects no longer extant.

Vegetable life, therefore, the object of proper agriculture, is the indispensable condition of at least the higher forms of animal existence, and the economical value of this art can hardly be overrated. But are its present condition and estimation answerable to its intrinsic importance? The result of agricultural labors depend upon causes so obscure, and difficult of appreciation, or, determined by meteoric and telluric influences apparently so completely beyond human foresight and control, or to express the same idea in a fewer words, nature here contributes so much and man so little, that the name of an art has been somewhat unwillingly accorded to the agricultural profession. But since modern analytical science has busied itself with economical investigations, agriculture has come to have its proper laws, and is no longer conducted by arbitrary rules, themselves founded on a blind and groping experience. Until chemical analysis had shown what were the constituent elements of vegetable forms, ascertained their proportions and modes of combination, and resolved soils into their primary ingredients, there was no apparent reason why a particular locality should be better suited to one vegetable growth than to another. The adaptation of a given crop to a given soil was simply an isolated fact, accidentally observed, and determined by no ascertained relation of cause and effect. But when the ingredients and their proportion, in wheat or any other grain, are known, we may at once infer what soils are best fitted for its production, as containing or supplying, in the truest measure, the elements which enter into that grain, and we are also taught at the same time by what means to bestow on particular soils the properties needed for the growth of particular vegetables. Thus agriculture acquires laws founded not on mere empiricism, but on established constant physical facts, and therefore becomes a branch of NATURAL KNOWLEDGE, instead of a mass of rules referable to no known general truth. The telluric influences to be regarded in agriculture are fast becoming entirely appreciable, and we shall doubtless soon know, with a close approximation to certainty, the relation between soils and crops, and the only merely empirical question remaining will be the economical inquiry (for the due solution of which, other and fluctuating elements are to be considered,) how far it is expedient to select crops adapted to particular localities, or by artificial means so to change the natural character of soils, as to fit them for the growth of given crops. So far then as telluric influences are concerned, it be may assumed, that the results of agricultural labors are in the main subject to calculation, and depend entirely upon the intelligence and industry of the husbandman.

But the equally important meteoric influences as yet elude our grasp. All the greater known causes on climate are constant, and therefore reasoning *a priori* we should be authorized to conclude, that the cycles of our seasons would be regular and invariable. The heavenly bodies whose movements occasion the alternation of spring and summer, and autumn and winter, revolve in almost unchanging orbits; the constituents of the atmosphere have been precisely determined and are everywhere and at all times substantially the same; here then are apparently sufficient elements of certainty, but the electrical, thermometrical and hygrometrical condition of the atmosphere, or in other words the distribution of electricity and light, heat and cold, moisture and drought, is controlled by causes which have hitherto baffled the researches of the acutest inquirers, and, in fact, so irregular, that the fickleness of wind and weather has passed into a proverb. In spite therefore of the ingenious speculations of Espy, which, however visionary, possess high interest, and by no means deserve the ridicule that sciolists and fools have heaped upon them, we have certainly as yet little cause to hope that climatic influences can ever be subject, in any important degree, to voluntary human modification or control. But though man cannot at his pleasure command the rain and the sunshine, the wind and frost and snow, yet it is certain that climate itself has in many instances been gradually changed and ameliorated or deteriorated by human action. The draining of swamps and the clearing of forests perceptibly effect the evaporation from the earth, and of course the mean quantity of moisture suspended in the air. The same causes modify the electrical condition of the atmosphere and the power of the surface to reflect, absorb and radiate the rays of the sun, and consequently influence the distribution of light and heat, and the force and direction of the winds. Within narrow limits too, domestic fires and artificial structures create and diffuse increased warmth, to an extent that may affect vegetation. The mean temperature of London is a degree or two higher than that of the surrounding country, and Palais believed, that the climate of even so thinly a peopled country as Russia was sensibly modified by similar causes.—But though, in general, climatic influences are beyond our reach, yet their pernicious tendencies may sometimes be neutralized or overcome. Every one must have observed, that the tender plants in gardens surrounded by high walls or buildings are more secure from frost than those in the open fields. A slight difference in the elevation, a judicious selection of exposure, a protection against cold winds by groves or fences, the accumulation of heat by rocks and stones or artificial walls, or its absorption in consequence of the dark color of the soil, irrigation, and lastly shelter and the application of artificial heat, may enable the skillful cultivator to grow plants, which properly belong to a climate many degrees nearer the equator than his own. Thus the orange and the lemon ripen in the open air in certain favored localities in the South of England. British conservatories will supply you with the native fruits of the West Indies fresh from the parent stalk, and Russian luxury has taught the flowers of the tropics to bloom in perpetual summer on the very verge of the frozen zone. Even the grape, so emphatically the child of the sun, is said to grow in England in greater perfection than in Madeira or the Levant, and the pineapples of European hot-houses surpass in weight and flavor the indigenous growth of the American Islands.

An interesting enquiry connected with this branch of our subject is that of the possibility of acclimating plants, or of so changing their habits, by a slow and gradual removal to a climate of lower temperature, that they will grow to perfection in a colder zone than nature seems to have designed them for. We do not perhaps know enough of the laws of vegetable life to be authorized *a priori* to determine this question, though eminent botanical physiologists, reasoning from acknowledged principles, have denied that any such change whatever can be effected. It is not very obvious why a change in this particular, which it would seem might be brought about without any appreciable transformation of organization, should be more impracticable than the apparently greater modifications, which we see cultivation every day produce in the habits and even structure of plants, and there are certainly some instances which, so far as those particular cases are concerned, seem to be completely successful experiments. So far as we are informed, no botanist has ever doubted or denied that all the numerous varieties of our common Indian corn are *specifically* one and the same plant; yet the seed which is grown within the tropics will not ripen in latitude forty-five. If you transport it even two degrees to the north, it is doubtful whether it will arrive at maturity; if you carry it four or five, it is certain that it will not. Yet by a very gradual process of successive slow removals from the South to the North, then there is no doubt that the maize of Carolina may be made to grow in Canada. The case of Indian corn is not an unique example of acclimation, though we are aware of no other instance equally striking and successful. The removal of a plant from a higher to a lower latitude is usually accompanied with an increase in size, and the annuals of the North sometimes become biennial or even perennial when transferred to a milder sky; but this latter change may be effected or reversed, at least in the bread stuff grains (spring and winter wheat, for example, being convertible into each other) without removal, and it appears to be in general true, that cultivated plants which grow through a wide range of latitude, undergo no modification from change of climate. The seed of wheat, which has been cultivated in Egypt ever since the reigns of the Pharaohs, or from the time of the conquest, on the table lands of Mexico, where it has but twelve hours of light and heat, will thrive as well in Canada, where the summer sun shines for sixteen hours, as if it has been brought over from Normandy by the fishermen of St. Malo. There are indeed varieties of the cerealia, or bread-stuff grains, as well as of almost all other vegetables that enter into the food of man, varying in their period of growth and time of ripening, but these often co-exist in the same latitude, and seem to be independent of climate. In general the grains of Vermont and Virginia, of the shores of the Mediterranean and those of the Baltic, thrive interchangeably, and the occasional increased rapidity of growth, which sometimes occurs upon transfer from a lower to a higher latitude, is due rather to a change of conditions than of habits, to an increase of stimulus from long continued light and heat, rather than to a modified organization.

Connected with this inquiry is another, of even greater interest, which is, how far can vegetables, without change of climate, be accelerated in growth, improved in quality, or increased in productiveness of fruit, by artificial means? It has been supposed that all domesticated plants were originally wild, and that they have been changed and improved by cultivation, through successive ages, and under different conditions, but this seems to be a hasty assumption. That certain vegetables, especially those of recent introduction into husbandry, have been considerably ameliorated, is known with reasonable certainty. The potato, when first brought to England, is described by contemporary writers as resembling the artichoke in size and consistence, and as being scarcely superior in flavor to that insipid root. We know too, that the average yield of many plants has been greatly increased, and that by the careful selection of seeds of accidental varieties, their maturity has often been considerably hastened, but to infer from these and other like facts the broad proposition, that all cultivated plants are wild vegetable reclaimed, ennobled, and, as it were, civilized by human art, is an unwarrantable generalization. What botanist has proved, upon scientific principles, the identity of wheat, or rye, or even oats, with any known plant of spontaneous growth? What experimental physiologist has succeeded in transforming any of the wild gramineous vegetables into the likeness of the cerealia, or has shown that these latter, if left to themselves, will go on self-reproductive, but degenerating at every remove, and finally relapsing into their supposed original condition of unfruitful grasses? Some botanists, indeed, identify the peach with the almond, and the apple with the medlar *[a cultivated European fruit]*, and it is confidently stated, that the native habitat of the potato has been not long since discovered, and the rudimental germ and primitive type of our Indian corn quite lately detected, in the wilds of South America. But we are not yet in possession of the requisite data for determining this interesting question, though the known fact, that wheat which has lain three thousand years in the catacombs of Egypt, will germinate and produce seed precisely resembling both itself and the grain now commonly grown in the same locality, certainly tends to prove that the cerealia of the old world undergo no change from long continued cultivation. I have already noticed that fact, that our only indigenous cereal grain, Indian corn, is an exception to the general rule, that these plants are unaffected by change of climate. Does this indicate an essential difference in character, or is it the result of its later reclamation, and consequently more incomplete adaptation to the uses of a migratory animal like man?

It is a characteristic distinction between domesticated and unreclaimed plants and animals, that the latter are much more constant in form and dimensions, and much less liable to what are called accidental varieties, than those whose original properties have been more or less modified by domestication. Unrestricted nature works with infinite variety of type, but in general in close conformity with her appointed models, and if the influences of cultivation are withdrawn for a few generations from a plant or an animal which has been transformed by domestication, it will resume its wild native uniformity of properties and shape. Wild plants of a given species, growing under similar circumstances, resemble each other so closely as hardly to be distinguishable, while the varieties of the same plant, when reclaimed are so numerous, and so aberrant both from each other and from the primitive type, that they seem to have nothing in common. The plumage of the wild turkey is uniformly black shot with bronze or gold, the domestic fowl struts in every variety of hue; the wild cattle of particular districts in South America, though we know them all to be descended from the variegated race and Old Spain are so exactly like each other, that the dealers can tell, simply by the shape of the horn, the region where they were bred. This tendency of domestication to multiply varieties, which may be easily rendered permanent, offers to man an inexhaustible field of improvement, in both vegetable and animal life, and the amelioration which the skill of English agriculturists has effected in this way are among the most remarkable triumphs of art over nature. How far this process can be carried is of course a question of experience, but there is every reason to suppose, that none of our cultivated vegetables have yet reached their highest point of attainable perfection, in regard either to quality or quantity of product. The increasing density of population, and the consequently enlarged demand for agricultural products in the old world, together with the spirit of rivalry and emulation in the new, will stimulate continued exertion in this honorable field of labor, and we may hope that the peaceful triumphs of the husbandman will eclipse in popular renown as well as in true utility, the proudest trophies of the warrior.

But besides improved modes of cultivation, we may look for great advantages to agriculture from the introduction of new vegetables and animals into our husbandry. If we remember how Europe and America have mutually enriched each other with new forms of productive life, and consider what vast provinces teeming with endless variety of vital existences still remain but half explored upon our own continent, if we reflect how little we know concerning the natural productions of the boundless realms of central Asia, and of China, where one third part of the human family subsists upon the fruits of rural industry, and where the art of the husbandman and gardener is said to have been carried to a pitch unknown in Europe, we may well imagine that our fields and fruityards and gardens are destined to acquire new sources of vegetable luxuriance and wealth and beauty from regions yet untrodden by Christian feet. But we have good cause to hope, that additions may be made to our stock of cultivated vegetables from more accessible sources. Compelled as we have been in this new world to avail ourselves of certain and approved methods of husbandry, we have not yet had time or means to experiment upon the economical value of wild indigenous plants, but since savage ingenuity has contrived to render innocuous the most poisonous vegetables, and to extract the nourishing tapioca from the deadly cassava, it seems not improbable that civilized chemistry may discover the elements of nutriment in many plants not hitherto used for human food. The native grasses of our country, too promise to prove highly valuable in agriculture, and the wild rice and wappatoo of our northern swamps, which are largely consumed by the Indian tribes, may perhaps prove worthy of cultivation, and thus give value to lands, which, in our present system of husbandry, are wholly unproductive.

There is not perhaps room for so sanguine expectations, in regard to increasing the variety of our domestic animals. We are probably acquainted with all the quadrupeds hitherto domesticated by man, and of these, it is certain that few not already introduced in European and American husbandry can be profitably added to the numbers we already possess. The powerful elephant requires the warm temperature and abundant vegetable food of the tropics; the buffalo possesses no superiority over our common black cattle, and we have no localities suited to the curiously peculiar structure of the camel, though it is possible that animal might thrive on the sands of the South or the great prairies of the Southwest, and there is strong reason to suppose, that the alpaca may be naturalized and reared with profit in many parts of Europe, as well as the American Union. The origin of our domestic fowls and quadrupeds is involved in the same obscurity as that of our cereal grains. Some of them are not known to exist, or to have ever existed, in a wild state, and we are at liberty, even apart from the evidence of scripture, to suppose that new-born man found himself at his first awakening surrounded by the grain of our fields, and the sheep of our pastures. The variety of tame fowls, on the other hand, might in all probability be considerably increased, and the success which has attended the domestication of the turkey, gives encouragement for trying the same experiment with other gallinaceous birds. It is obvious, however, that the chief improvements in husbandry are to be expected from the continued application of natural science to the resolution of yet undetermined problems in vegetable physiology, and from the employment of new agents as stimulants of growth. Among these, are various mineral substances and chemical preparations and not least, the electric fluid, which is known to exercise a powerful influence upon vegetation, though under conditions too obscure to be yet well appreciated. In these researches, the man of science must precede the operative farmer, and theories must be digested in the closet, before they are reduced to practice in the field. But in the mean time, there is abundant room for improvement in the use of means already known and familiar. I may mention the better economy of manures, and particularly, the saving, for this purpose, of the various highly fertilizing substances which are produced in our common household operations; the introduction of improved agricultural implements; the drainage of the soil, which not only restores waste lands to agriculture, but is an important means of securing the farmer against the great enemy of his crops, the frosts of autumn and spring; the extirpation of thistles and other weeds, and the destruction of noxious insects; the boring of artesian wells for supplying our dry pastures; the use of hedges on some soils, and even of iron in other localities, instead of our common costly and perishable wooden fences, and especially the substitution of cheap mechanical power of water and of wind for manual labor.

It is little to the credit of our agriculturists, that the greatest progress in these and other modern improvements should have been made by persons not bred to agricultural pursuits, and it has often been said that mechanics, merchants and professional men make in the end the best farmers. If there is any truth in this opinion, it is probably because these persons, commencing their new calling at a period of life when judgment is mature, tied down by habit to no blind routine of antiquated practice, and ridden by no nightmare of hereditary prejudice in regard to particular modes of cultivation, are conscious of the necessity of observation and reflection, in an occupation, the successful pursuit of which requires so much of both, and feel themselves at liberty to select such processes as are commended by the results of actual experience, or accord with the known laws of vegetable physiology. Under such circumstances, a judicious man, encouraged by the stimulus of novelty, would be likely to study the subject with earnestness, and to profit by his own errors, as well as by the experience of others.

There are certain other improvements connected with agriculture, to which I desire to draw your special attention. One of these is the introduction of a better economy in the management of our forest lands. The increasing value of timber and fuel ought to teach us, that trees are no longer what they were in our fathers' time, an incumbrance. We have undoubtedly already a larger proportion of cleared land in Vermont than would be required, with proper culture, for the support of a much greater population than we now possess, and every additional acre both lessens our means for thorough husbandry, by disproportionately extending its area, and deprives succeeding generations of what, though comparatively worthless to us, would be of great value to them. The functions of the forest, besides supplying timber and fuel, are very various. The conducting powers of trees render them highly useful in restoring the disturbed equilibrium of the electric fluid, they are of great value in sheltering and protecting more tender vegetables against the destructive effects of bleak or parching winds, and the annual deposit of the foliage of deciduous trees, and the decomposition of their decaying trunks, form an accumulation of vegetable mould, which gives the greatest fertility to the often originally barren soils on which they grow, and enriches lower grounds by the wash from rains and the melting of snows. The inconveniences resulting from a want of foresight in the economy of the forest are already severely felt in many parts of New England, and even in some of the older towns in Vermont. Steep hill sides and rocky ledges are well suited to the permanent growth of wood, but when in the rage for improvement they are improvidently stripped of this protection, the action of sun and wind and rain soon deprives them of their thin coating of vegetable mould, and this, when exhausted, cannot be restored by ordinary husbandry. They remain therefore barren and unsightly blots, producing neither grain nor grass, and yielding no crop but a harvest of noxious weeds, to infest with their scattered seeds the richer arable grounds below. But this is by no means the only evil resulting from the injudicious destruction of the woods. Forests serve as reservoirs and equalizers of humidity. In wet seasons, the decayed leaves and spongy soil of woodlands retain a large proportion of the falling rains, and give back the moisture in time of drought, by evaporation or through the medium of springs. They thus both check the sudden flow of water from the surface into the streams and low grounds, and prevent the droughts of summer from parching our pastures and drying up the rivulets which water them. On the other hand, where too large a proportion of the surface is bared of wood, the action of the summer sun and wind scorches the hills which are no longer shaded or sheltered by trees, the springs and rivulets that found their supply in the bibulous soil of the forest disappear, and the farmer is obliged to surrender his meadows to his cattle, which can no longer find food in his pastures, and sometime even to drive them miles for water. Again, the vernal and autumnal rains, and the melting snows of winter, no longer intercepted and absorbed by the leaves or the open soil of the woods, but falling everywhere upon a comparatively hard and even surface, flow swiftly over the smooth ground, washing away the vegetable mould as they seek their natural outlets, fill every ravine with a torrent, and convert every river into an ocean. The suddenness and violence of our freshets increases in proportion as the soil is cleared; bridges are washed away, meadows swept of their crops and fences, and covered with barren sand, or themselves abraded by the fury of the current, and there is reason to fear the valleys of many of our streams will soon be converted from smiling meadows into broad wastes of shingle and gravel and pebbles, deserts in summer, and seas in autumn and spring. The changes, which these causes have wrought in the physical geography of Vermont, within a single generation, are too striking to have escaped the attention of any observing person, and every middle-aged man who revisits his birth-place after a few years of absence, looks upon another landscape than that which formed the theatre of his youthful toils and pleasures. The signs of artificial improvement are mingled with the tokens of improvident waste, and the bald and barren hills, the dry beds of the smaller streams, the ravines furrowed out by the torrents of spring, and the diminished thread of interval that skirts the widened channel of the rivers, seem sad substitutes for the pleasant groves and brooks and broad meadows of his ancient paternal domain. If the present value of timber and land will not justify the artificial re-planting of grounds injudiciously cleared, at least nature ought to be allowed to reclothe them with a spontaneous growth of wood, and in our future husbandry a more careful selection should be made of land for permanent improvement. It has long been a practice in many parts of Europe, as well as in our older settlements, to cut the forests reserved for timber and fuel at stated intervals. It is quite time that this practice should be introduced among us. After the first felling of the original forest it is indeed a long time before its place is supplied, because the roots of old and fell grown trees seldom throw up shoots, but when the second growth is once established, it may be cut with great advantage, at period of about twenty-five years, and yields a material, in every respect but size, far superior to the wood of the primitive tree. In many European countries, the economy of the forest is regulated by law; but here, where public opinion determines, or rather in practice constitutes law, we can only appeal to an enlightened self-interest to introduce the reforms, check the abuses, and preserve us from an increase of the evils I have mentioned.

There is a branch of rural industry hitherto not much attended to among us, but to the social and economical importance of which we are beginning to be somewhat awake. I refer to the agreeable and profitable art of horticulture. The neglect of this art is probably to be ascribed to the opinion, that the products of the garden and the fruityard are to be regarded rather as condiments or garnishings than as nutritious food, as something calculated to tickle the palate, not to strengthen the system; as belonging in short to the department of ornament, not to that of utility. This is an unfortunate error. The tendency of our cold climate is to create an inordinate appetite for animal food, and we habitually consume much too large a proportion of that stimulating aliment. This, when we compare the relative cost of a given quantity of nutritive matter obtained from animals and vegetables, seems very indifferent economy, and considerations of health most clearly indicate the expediency of increasing the proportion of our fruit and vegetable diet. We cannot in this latitude expect to rival the pomona of more favored climes, but in most situations, we may, with little labor or expense, rear such a variety of fruits as to supply our tables with a succession of delicious and healthful viands throughout the entire year. It is for us a happy circumstance, that most fruits attain their highest perfection near the northern limit of their growth, and though the fig and the peach cannot be naturalized among us, we may, to say nothing of the smaller fruits, successfully cultivate the finer varieties of the apple, the pear and even the grape.

Another mode of rural improvement may be fitly mentioned in connection with this last. I refer to the introduction of a better style of domestic architecture, which shall combine convenience, warmth, and reasonable embellishment. A well arranged and well proportioned building costs no more than a misshapen disjointed structure, and commodity and comfort may be had at as cheap a rate as inconvenience and confusion. Neither is a little expenditure in ornament thrown away. The paint which embellishes tends also to preserve, and the shade trees not only furnish a protection against the exhausting heats of summer, but they serve, if thickly planted, to break the fury of the blasts of winter, and in the end they furnish a better material for fuel or mechanical uses than the spontaneous forest growth. The habit of domestic order, comfort, and neatness will be found to have a very favorable influence in the manner in which the outdoor operations of husbandry are connected. A farmer, whose house is neatly and tastefully constructed and arranged, will never be a slovenly agriculturist. The order of his dwelling and his courtyard will extend to his stables, his barns, his graneries, and his fields. His beasts will be well lodged and cared for, his meadows free from stumps, and briars, and bushes, and the strength of his fences will secure him against the trespasses of his thriftless neighbor's unruly cattle. Another consideration, which most strongly recommends attention to order and comfort and beauty in domestic and rural arrangements is that all these tend to foster a sentiment, of which the enterprising and adventurous Yankee has in general, far too little—I mean a feeling of attachment to his home, and by a natural association, to the institutions of his native New England. To make our homes in themselves desirable is the most effectual means of compensating for that rude climate which gives us three winters each year—two Southern, with a Siberian interrelated between—and of arming our children against the tempting attractions of the milder sky and less laborious life of the South, and the seductions of the boasted greatness and exaggerated fertility of the West. A son of Vermont who has enjoyed, beneath the paternal roof the blessings of a comfortable and well ordered home, and whose eye has been trained to appreciate the charms of rural beauty, which his own hands have helped perhaps to embellish, will find little to please in the slovenly husbandry, the rickety dwellings, and the wasteful economy of the Southern planter, little to admire in the tame monotony of a boundless prairie, and little to entice in the rude domestic arrangements, the coarse fare and the coarser manners of the Western squatter. A youth will not readily abandon the orchard he has dressed, the flowering shrubs which he has aided his sisters to rear, the fruit or shade tree planted on the day of his birth, and whose thrifty growth he has regarded with as much pride as his own increase of stature and who that has been taught to gaze with admiring eye on the unrivalled landscapes unfolded from our every hill, where lake, and island, and mountain and rock, and well-tilled field, and evergreen wood, and purling brook, and cheerful home of man are presented at due distance and in fairest proportion, would exchange such scenes as these, for the mirey sloughs, the puny groves, the slimy streams, which alone diversify the dead uniformity of Wisconsin and Illinois!

I have now shown, I hope, rather by suggestion than by argument, that the profession of agriculture in this age and land is an honorable, and in its true spirit, an elevated and an enlightened calling. I have adverted to its importance as an instrument of primary civilization, endeavored to indicate its present position as an art, and hinted at its future hopes and encouragements. It only remains for me to say a word on that other great branch of industry, the promotion of which is one of your leading objects, the ARTS OF

CONVERSION, namely, or as they are more generally called, the MECHANIC ARTS.

Although these arts are practiced to some extent by the rudest savages, as I said at the outset, yet they do not in general attain to any considerable degree of perfection, until agriculture has made great advances, and as they are the last of the industrial arts to be fully developed, so are they the ultimate material means, by which the power, and wealth, and refinement of social man are carried to their highest pitch. The distinction between agriculture and proper mechanic art may be thus stated. The one avails itself of the organic forces of nature, for the purpose of simple reproduction and multiplication; the other employs the more powerful inorganic forces for the conversion of natural forms into artificial shapes. The mechanic derives the raw material directly from the hand of nature, but the form, character and properties of the final product are determined by human contrivance, sometime relying upon the plastic power of the hand, and at other times aided by natural forces, which man has learned to guide and control. Out of a mass of iron, the artisan can forge at his pleasure, a sword or a plough-share. He can fashion from a block of wood a spinning wheel or a heathen idol. With a flask of mercury, he can silver a mirror, supply a hospital with a month's stock of calomel, or extract from ore and dross an ingot of gold. He can coin that same gold into Republican eagles, or royal sovereigns, gild with it the dome of a royal mosque or a capitol, or draw it out into a wire and twist it into a lady's necklace. He can convert a bale of cotton into muslin that shall rival the fineness of the spider's web, canvas and cordage for a ship of the line, or an explosive substance, an ounce of which shall rend that ship into ten thousand fragments. He can hew from the dead marble a chimney piece for a palace or a cottage, the mausoleum of a Napoleon, a baptismal font, or the speaking statue of a Washington. The whole art of the agriculturist on the other hand is exhausted in the multiplication of certain natural products all substantially the original. Here the seed from the storehouse of the sower becomes as it were the raw material, or rather the model, and nature is the artificer, whom man compels to repeat and reproduce the works of her own mysterious cunning. The labors of the agriculturist are confined to the production and slight improvement of the comparatively few natural forms which he has learned to make subservient to his own uses; the toils and objects of the mechanic are as diversified as the wants and the inventive capacity of man.

But between these two great branches of productive industry, diverse as are their objects and their processes, there is neither interference nor competition; each depends upon and is in turn helpful to the other, and prosperity has crowned no country, which has adapted its legislation exclusively to the encouragement of either. As a general rule, it may be said, that mechanical operations absorb a much larger amount of agricultural products than mere agricultural consumers of the results of mechanical labor, and therefore the husbandman is directly interested in the prosperity of the mechanic, who is his best customer. A single operative in almost any branch of mechanical art works up a vastly greater amount of raw material than one agricultural laborer can grow, and produces a much larger quantity of the manufactured article than one laborer can consume. Three thousand spinners, weavers, dyers and finishers, will convert into dressed cloth all the wool grown in Vermont, and the cloth they will produce would furnish two full suits a year to every male inhabitant of the State. In the case of cottons, the manufacture of which is simple, and requires less manipulation, the disproportion between the quantity of raw material produced by one field laborer and consumed by one manufacturer is even greater; and a similar rule holds true, in general, of all the mechanic arts. The obvious reason of this disproportion between the results of labor is, that the mechanic performs the heaviest portion of his work by mechanical contrivances, which press into his service the inexhaustible inorganic forces of nature, and enable a single individual to wield more than the strength of a thousand, while the agriculturist accomplishes his task by mere bone and sinew, the unaided force of man and beast, and looks to the comparatively feeble and uncertain powers of organized nature to bring about the wished for result. There is another reason why the mechanic arts are of great and perpetually increasing value to the agriculturist. They are constantly discovering new uses, and thereby extending the demand for the raw material. Who can compute the increased value that the invention of paper has given to vegetable fibrous substance, or that of explosive properties to cotton? How many acres has the use of starch in manufacturing added to the culture of the potato,and how many field laborers find employment in growing madder and the teazle? The farmer is interested in the prosperity of the workshop, because it offers a market for his raw material and his surplus food, furnishes occupation in mechanical employments, and thereby reduces the number of rival laborers in agriculture, and cheaply supplies him with wares and implements, which must otherwise be, in the words of the homely proverb, "far fetched and dear bought."

But besides these considerations, there are other reasons of a higher character, why the farmer, in common with all wise and good citizens, should esteem the mechanic arts in an eminent degree worthy of patronage and encouragement, especially in a country whose yet undeveloped natural resources are so boundless and diversified. It is by means of these arts alone, that those internal improvements can be effected, which bind our wide confederacy together, unite our inland seas with the ocean—the common boundary and highway of nations, bring every producer within reach of a market, and tend to equalize the value of lands in all parts of our wide domain. On these topics I am sure I speak to neither ignorant nor uninterested ears; and the efforts which the people of Vermont are now making to secure to themselves the advantages of the improvements to which I allude, are as creditable to their intelligence, as they are honorable to their public spirit. In the distribution of the bounty of the national government, the power of the larger states will never allow to our smaller commonwealths their just share, and whatever millions may be lavished on the favored West, we must be content with such improvements as the means accumulated by our own industry, aided by the enlightened liberality of our city capitalists, shall enable us to make. It is therefore a highly encouraging cause of hope and satisfaction that Vermont has at length put her own shoulder to the wheel, without waiting for Hercules, and we have every reason to expect that another year or two will place our territory on as favorable a ground as localities more blessed by nature, or more pampered by the partial bounty of the general government. What great advantage over you in fact has the farmer who lives on a turnpike road ten miles from Boston, if the grain which was removed from your storehouse at sunset is, while he sleeps, hurried to market on the wings of steam, and delivered at the city depot before he has time to transport thither the corn he had measured and loaded up before he retired to rest? And why will not our water-power be as available as that of Massachusetts, when the cottons turned out by our factories today shall be afloat on the Atlantic tomorrow?

It is through the mechanic arts alone, that we can become truly independent of foreign nations, and establish an interchange between the producer, the manufacturer, and the consumer, which shall increase the wealth and lighten the burdens of each, by retaining among ourselves the net profits of labor, and thus avoiding the drains of the precious metals for supplies. The mechanic arts are worthy of patronage from their progressive character, and the promise they hold out to us of acquiring a complete mastery over inanimate nature. The progress of agriculture, within the last half century though great in itself and full of future promise, has been but a tardy movement, in comparison with the swift advancement of the mechanic arts. The steamboat, the locomotive, the power loom, and the power press, have all been brought into use since the beginning of the present century, and what a revolution have they wrought upon the face of the globe! How they have brought together and linked different states and countries! What millions have they clothed, and what millions enlightened! Suppose we were at once to be deprived of these great gifts of mechanic art, and suddenly cut off from the cheap and abundant supply of the means of knowledge, our necessary clothing doubled in cost, and our products reduced to half their value for want of speedy and economical means of transport to their market, our intelligence from the seat of government, from our distant friends, and from the old world, as well as our personal communication with other parts of our country, retarded and delayed for want of our accustomed means of transport and locomotion, what value should we not attach to these now almost unnoticed blessings, and what efforts and sacrifices should we not be ready to encounter to regain them? Yet we may well judge of the future from the past, and the progress of natural knowledge, upon which all mechanical art is founded, authorizes us to expect the remaining half of the nineteenth century will be as fertile in improvements as the portion of it which has already elapsed. The mechanic arts are eminently democratic in their tendency. They popularize knowledge, they cheapen and diffuse the comforts and elegancies as well as the necessaries of life, they demand and develop intelligence in those who pursue them, they are at once the most profitable customers of the agriculturist, and the most munificent patrons of the investigator of nature's laws.

Thus, then, the several branches of productive industry, for the promotion of which you are associated, mutually cherish and depend upon each other. The herdsman, the ploughman and the mechanic are fellow laborers, not indeed competitors, but co-workers in a common cause, and every measure that tends to elevate any one of them at the expense of another, must in the end infallibly prove detrimental to the best interests of them all.

Lectures Delivered before the Smithsonian Institution No. I—The Camel

[1855; published in the Ninth Annual Report of the Smithsonian Institution for 1854, pp. 98 – 122, U.S. Senate, 33rd Congress, 2nd session, misc. doc. 24]

Marsh often contradicted himself in his own writings, sometimes making it difficult to know how deeply he believed his own conclusions about the dangers posed by human modification of the landscape and organic world. In this 1854 lecture for the Smithsonian Institution, published in 1855 (and expanded into a book in 1856), Marsh strongly advocated the introduction of camels, native only to the Old World regions of Asia and northern Africa, into the southwestern part of the United States for use by the military. This lecture, which he based on a number of his informal writings on this subject during the preceding few years, does not reveal any hesitation on his part about the advisability of such a manipulation of a region's fauna, and he went to great lengths to develop a convincing argument and an introduction strategy that he felt had the greatest chance of success.

Yet only eight years later, Marsh demonstrated that he was aware of the problems posed by the introduction of exotic animals (*Man and Nature*, chapter 2), and his fervor for the introduction of the camel into the United States seems paradoxical. The best explanation for Marsh's position, beyond the acknowledgment that his thinking may have evolved between the time he wrote this lecture and *Man and Nature*, lies in the recognition that he was extremely supportive of the agricultural arts and sciences in all their forms. In fact, Marsh can fairly be characterized as having a greater concern throughout his life for the future of agriculture and agriculturalists than for conservation and environmental protection. He felt that the domestication of plants and animals was not only appropriate but desirable; to Marsh, the practice of domestication was part of humanity's great Christian mission on Earth and a fundamental characteristic of higher civilization. Thus, his advocacy for introducing the camel was less about the introduction of an exotic herbivore and more about the importation and exploitation of a domesticated quadruped that would be kept under human control at all times and serve the needs of society. Support for this interpretation is found in the fact that in *Man and Nature* all of Marsh's examples for the negative consequences of exotic species are drawn from insects, worms, and plants.

The actual history of the camel in the United States, however, demonstrates how difficult it can be to maintain absolute control over domesticated animals. Based on the urging of a few officers, the Army did, in fact, import camels for use by its cavalry in the Southwest. In 1856, a year after Marsh's lecture was published, thirty-four camels, purchased in Egypt by agents for the U.S. government, arrived in Indianola, Texas, and became the foundation of the U.S. Camel Corps. The camels were poorly understood by most of the men assigned to the Corps, and were little used except during an exploration of routes between Texas and California in 1857. After the Civil War, the Corps was disbanded and the camels were sold to various people, some of whom released their animals into the wild. Fortunately, unlike similar introductions of horses, donkeys, goats, and pigs, this introduction of the camel into the wild failed, and although camels were still sighted in the desert through the early 1900s, they did not successfully reproduce and no feral camels currently roam the deserts of North America.

The reason for the failure is not known. As Marsh pointed out in this lecture, the camel is well suited for the deserts of North America and should have faced no difficulty finding food or tolerating the weather. Studies over the last several decades on the biology of species introductions have shown that the probability of an introduction being successful increases with both the number of individuals introduced at a single time and the number of times an introduction is attempted. Perhaps the only reason camels did not establish themselves following their release was that so few individuals were involved and the releases were so widely scattered that breeding pairs never had a chance to form. For this we can be grateful; since the beginnings of transoceanic explorations in the fifteenth century, many species of mammalian herbivores have been introduced widely on continents and islands around the world—especially goats, pigs, and rabbits—and through their destruction of plant populations and competition with native herbivores they constitute one of the greatest threats to ecological health today.

The first command addressed to man by his Creator, and substantially repeated to the second great progenitor of our race, not only charged him to subdue the earth, but gave him dominion over all terrestrial creatures, whether animate or inanimate, and thus predicted and prescribed the subjugation of the entire organic and inorganic world to human control and human use.

Man is yet far from having achieved the fulfilment of this grand mission. He has, indeed, surveyed the greater part of his vast domain; marked the outline of its solid and its fluid surface, and approximately measured their areas and determined their relative elevation; pierced its superficial strata, and detected the order of their historical succession; reduced to their primal elements its rocks, its soils, its waters, and its atmosphere, and even soared above its canopy of cloud. He has traced, through the void of space, its movements of rotation, revolution, and translation; resolved the seeming circles of its attendant satellite into strangely tortuous paths of progression; investigated its relations of density, attraction, and motion, to other visible and invisible cosmical orbs; and unfolded the laws of those mysterious allied agencies, heat, light, electricity, and magnetism, whose sphere of influence seems commensurate with that of creation. But, notwithstanding these triumphs, earth is not yet all his own; and millions of leagues of her surface still lie uninhabited, unenjoyed, and unsubdued—yielding neither food, nor clothing, nor shelter to man, or even to the humbler tribes of animal or vegetable life, which, in other ways, minister to his necessities or his convenience.

In like manner, man has studied the biography, and the relations of affinity or dependence, of the infinitely varied contemporaneous forms of organic life; traced the history of myriads of species of both plants and animals, which had ceased to be before the Creator breathed into *his* nostrils the breath of life; and demonstrated the past and present existence of numerous tribes of organic beings, too minute to be individually cognizable by any of the unaided senses, and yet largely influencing our own animal economy, and even composing no unimportant part of the crust of the solid globe; but of the vegetables that clothe and diversify its soil, of the animated creatures that float in its atmosphere, enliven its surface, or cleave its waters, but comparatively few have as yet been rendered in any way subservient to human use, fewer still domesticated and made the permanent and regular denizens of his fields or companions of his household.

The efforts of civilized man towards the fulfilment of this great command have been directed almost exclusively to the conquest of inorganic nature, by the utilization of minerals; by contriving methods for availing himself of the mechanical powers and of natural forces, simply or in cunning combinations; by cutting narrow paths for facilitating travel and transport between distant regions; and by devising means of traversing with certainty and speed the trackless and troubled ocean.

The proper savage smelts no ores, and employs those metals only which natural processes have reduced. He binds the blocks of which he rears his temples with no cement of artificial stone. He drains no swamps, cuts no roads, excavates no canals, turns no mills by power of water or wind, and asks from inorganic nature no other gifts than those which she spontaneously offers, to supply his wants and multiply his enjoyments.

On the other hand, the very dawn of social life, in those stages of human existence which quite precede all true civilization, demands, as an indispensable condition, not the mere usufruct of the spontaneous productions of organic nature, but the complete appropriation and domestication of many species of both plants and animals. Man begins by subjugating, and thereby preserving, those organic forms which are at once best suited to satisfy his natural wants, and, like himself, least fitted for a self-sustaining, independent existence (It is not the domestic animals alone whose existence is perpetuated by the protective, though often unconscious, agency of man. In the depths of our northern forests the voice of the song-bird, or of the smaller quadrupeds, is but seldom heard. It is in the fields tilled by human husbandry that they find the most abundant nutriment, and the surest retreat from bird and beast of prey. The vast flights of the wild pigeon are found, not in the remote, primitive woodlands, but along the borders of the pioneer settlements; and, upon our western frontier, it is observed that the deer often multiply for a time after the coming in of the whites, because the civilized huntsman destroys or scares away the wolf, the great natural enemy of the weaker quadrupeds.); and he is to end by extending his conquests over the more widely dissimilar, remote, and refractory products of creative nature. We accordingly owe to our primeval, untutored ancestors, the discovery, the domestication, the acclimation of our cereal grains, our edible roots, our improved fruits, as well as the subjugation of our domestic animals; while civilized man has scarcely reclaimed a plant of spontaneous growth, or added a newly tamed animal to the flocks and herds of the pastoral ages. Indeed, so remote is the period to which these noble triumphs of intelligent humanity over brute and vegetable nature belong, that we know not their history or their epochs; and if we believe them to be in fact human conquests, and not rather special birth-day gifts from the hand of the Creator, we must admit that cultivation and domestication have so completely metamorphosed and diversified the forms and products, and modified the habits, and even, so to speak, the inborn instincts of both vegetables and animals, that but the fewest of our household beasts and our familiar plants can be certainly identified with the primitive stock. Most of these, it is probable, no longer occur in their wild state and original form; and it is questionable whether they are even capable of continued existence without the fostering care of man.

In both these great divisions of organic life there are some species peculiarly suited to the uses of man as a migratory animal. The bread stuffs of the old world, and, in a less degree, our only American cereal, Indian corn, the pulse, the cucurbitaceous plants, and the edible roots of our gardens, as well as the horse, the dog, the sheep, and the swine, seem almost exempted from subjection to climatic laws. While, therefore, a degree of latitude, a few hundred feet of elevation, a trifling difference in soil, or in the amount of atmospheric humidity, oppose impassable barriers to the diffusion of most wild plants and animals, the domesticated species I have enumerated follow man in his widest wanderings, and make his resting-place their home, whether he dwells on a continent or an island, at the level of the sea or on the margin of Alpine snows, beneath the equator or among the frosts of the polar circle.

Others, again, of the domesticated families of the organic world seem, like the untamed tribes, inexorably confined within prescribed geographical bounds, and incapable of propagation or growth beyond their original limits; while others still, though comparatively independent of climate and of soil, are nevertheless so specially fitted to certain conditions of surface, and certain modes of human life, to the maintenance of which they are themselves indispensable, that even the infidel finds, in these mutual adaptations, proofs of the existence and beneficent agency of a self-conscious and intelligent creative power.

Among the animated organisms of this latter class, the camel is, doubtless, the most important and remarkable. The Ship of the Desert has navigated the pathless sand-oceans of Gobi and the Sahara, and thus not only extended the humanizing influences of commerce and civilization alike over the naked and barbarous African and the fur-clad Siberian savage, but, by discovering the hidden wells of the waste and the islands of verdure that surround them, has made permanently habitable vast regions not otherwise penetrable by man. The "howling wilderness" now harbors and nourishes numerous tribes in more or less advanced stages of culture; and the services of that quadruped, on which Rebekah journeyed to meet her spouse, and which, though neglected and despised by the polished Egyptian, constituted a principal item in the rural wealth of the father of Joseph, are as indispensable to these races, as are those of any other animal to man in any condition of society.

The camel lives and thrives in the tropics; through almost the whole breadth of the northern temperate zone; and is even met beyond Lake Baikal in conjunction with the reindeer, with which, among some of the northern tribes, he has exchanged offices, the deer serving as a beast of the saddle, while the camel is employed only for draught or burden. But his appropriate home is the desert, and it is here alone that he acquires his true significance and value, his remarkable powers being the necessary condition and sole means by which man has in any degree extended his dominion over the Libyan and the Arabian wildernesses.

In presence of the improvements of more advanced stages of society, the camel diminishes in numbers and finally gives place to other animals better suited to the wants and the caprices of higher civilization. Upon good roads, other beasts of draught and burden are upon the whole more serviceable, or, to speak more accurately, more acceptable to the tastes of cultivated nations; and the ungainly camel shares in the contempt with which the humble ass, the mule, and even the ox, are regarded by the polished and the proud. Besides this, both the products and the restraints of proper agriculture are unfavorable to his full development and physical perfection. When the soil is enclosed and subjugated, and the coarse herbage and shrubbery of spontaneous growth are superseded by artificial vegetation, he misses the pungent and aromatic juices which flavor the sun-burnt grasses and wild arborescent plants that form his accustomed and appropriate diet; the confinement of fence, and hedge, and stall are repugnant to his roving propensities and prejudical to his health, and he is as much out of place in civilized life as the Bedouin or the Tartar. Hence the attempts to introduce him into Spain, Italy, and other European countries have either wholly failed, or met with very indifferent success; and though he still abounds in Bessarabia, the Crimea, and all the southeastern provinces of Russia, yet the rural improvements which the German colonists have introduced into those regions have tended to reduce his numbers. When the wandering Tartar becomes stationary, encloses his possessions, and converts the desert steppe into arable ground, his camels retreat before the horse, the ox, and the sheep, and retire to the wastes beyond the Don and the Volga. So essentially nomade indeed is the camel in his habits, that the Arab himself dismisses him as soon as he acquires a fixed habitation. The oases of the desert are generally without this animal, and he is not possessed by the Fellaheen of the Sinaitic peninsula, by the inhabitants of Sinah or the oasis of Jupiter Ammon, or by those who cultivate the valleys of Mount Seir.

Of the primitive races of man, known to ancient sacred and profane history, but one, the Bedouin Arab, has retained unchanged his original mode of life. It is the camel alone, whose remarkable properties, by making habitable by man regions inaccessible to the improvements of civilization, has preserved to our own times that second act of the great drama of social life, the patriarchal condition. The Arab in all his changes of faith, heathen, christian, mussulman, has remained himself immutable; and the student of biblical antiquity must thank the camel for the lively illustrations of scripture history presented by the camp of the Ishmaelite sheikh, who is proud of his kindred with the patient Job, and who boasts himself the lineal descendant of Ibrahim el Khaleel, or Abraham "the friend" of God.

. . . The question of the practicability and advantages of introducing the camel into the United States for military and other purposes, is one of much interest and importance; and I hope I shall be pardoned if I prolong a discourse, which I fear has proved but a dry one, for the sake of suggesting some considerations upon a topic to which I have devoted some attention, both at home and abroad.

Among those who are practically familiar with the habits and properties of the camel, and who have studied the physical conditions of our territory west of the Mississippi, there is, I believe, little or no difference of opinion on the subject; and I am persuaded that the ultimate success of judicious and persevering effort is certain, and will be attended with most important advantages. At the same time, it must not be concealed that, as much depends on a point that nothing but experience can determine,—the selection, namely, of the particular breeds best adapted to our climate, soil, and other local conditions,—the result of a first experiment, unless tried on a liberal scale, and with animals of more than a single variety, is extremely uncertain. The question must be considered under two aspects: the one regarding the camel as simply a beast of burden; the other, his value as an animal of war. But even if it is conceded, which I by no means admit, that the organization of a proper mounted dromedary corps is impracticable or inexpedient, it does not, by any means, follow that the camel may not be of great value in the commissariat, and in all that belongs to the mere movement of bodies of men, as well as in the independent transportation of military stores and all the munitions of war.

The first question to be discussed is the adaptation of any variety of either species to the climate and soil of any portion of our territory. So far as mere extremes of temperature are concerned, it is quite certain that we have nowhere, west of the Mississippi, fiercer or more long-continued heats, more parched deserts, or wastes more destitute of vegetation, than those of the regions where the Arabian camel is found in his highest perfection; and the Bactrian thrives in climates as severe as even the coldest portion of our northeastern territory.

. . . So far, then, as climate and soil are concerned, it may be regarded as quite certain that the Bactrian camel can sustain any exposure to which he would be subjected in our trans-Mississippian territory; and there is no reason to doubt that the mezquit, acacia, and other shrubs, and the saline plants known to exist in many of those regions, would furnish him an appropriate and acceptable nutriment.

I cannot speak with equal confidence of the ability of the Arabian camel, and especially of the maherry of the desert, to bear corresponding trials. All high-bred animals are delicate, and impatient of exposure to great extremes and sudden changes; and although Denham and Clapperton speak of hard frosts in latitude 13 degrees north, and Lyon records a temperature four degrees below the freezing point, in districts constantly traversed by the maherry, yet the finest and fleetest animals will not bear the winter climate of Algiers. But, although we may not be able to breed dromedaries of a speed equal to the most extraordinary performances I have described, there is no reason to doubt that the more common animal, which will travel eight or ten hours a day at five miles an hour, for many days in succession, and with greater speed for a shorter period, can be bred and used with advantage throughout our southwestern territories, and on all the more southern passes of the mountains which divide the valley of the Mississippi from the Pacific slope, as well as throughout the State of California.

The ancient Asiatics, and, at a later period, the Romans, made a very extensive use of the dromedary in war, not only for the transportation of men and munitions, but as technical cavalry in actual combat; and they are still employed in Persia, Bokhara, and Tartary, for military purposes, and especially for the conveyance of light pieces of artillery, which are mounted between the humps, and used in that position, the camel kneeling while the gun is loaded, aimed and fired. In modern European armies they have hardly been employed, except by Napoleon, in transporting the baggage of his army in the Syrian campaign, in his celebrated dromedary regiment; and, more recently, by the army of occupation in Algeria. Upon the march from Egypt to Syria, the baggage, the camp equipage, and the sick, of an army of 15,000 men, were transported solely by camels.

It is remarkable that the military archives of France furnish little or no information, beyond the mere number of the corps, respecting the dromedary regiment of the army of Egypt, the historical documents belonging to the subject having been chiefly lost or suppressed; and all we know concerning it is derived from an imperfect and erroneous account in the great work on Egypt, and a late paper by Gourard, one of the savants who accompanied the expedition. Without entering into minute detail, it must suffice to say that this regiment, which numbered something less than 500 men, was organized in the main like a regiment of cavalry, and performed the same general service, with the most brilliant success. Although the men were taken from the infantry, a very short time was required to teach them the new discipline and drill, and the animals were habituated to the necessary evolutions in an incredibly short space of time. The services rendered by the corps were of a most important character, and its performances, according to Prétat, were quite unprecedented in military annals. This officer states that the ordinary march of the regiment was thirty French leagues, or about seventy-five miles, *without a halt;* and that a detachment belonging to it marched six hundred miles in eight days. These latter extraordinary statements rest on the testimony of a single individual, and though the corps was composed wholly of picked animals and picked men, and animated by the energy of a Bonaparte, it is very difficult to yield them full credence.

The experiments in Algeria, though satisfactory to the officers charged with them, whose reports seem entirely conclusive upon the value and economy of the camel as an animal of war, have been attended with less brilliant results. The prejudices of the officers and men against the use of this awkward and ungraceful animal in the regular service have proved very difficult to overcome. The peculiar organization of the French commissariat has interposed serious pecuniary obstacles, and the government has always seemed disinclined to consider this question in a spirit of liberality and candor. It is, however, proved that the use of the dromedary contributes in a most important degree to the economy, the celerity, and the efficiency of military movements in desert regions; and I cannot doubt that it would prove a most powerful auxiliary in all measures tending to keep in check the hostile Indians on the frontier, as well as in maintaining the military and postal communication between our Pacific territory and the east.

There are few more imposing spectacles than a body of armed men advancing under the quick pace of the trained dromedary; and this sight, with the ability of the animal to climb ascents impracticable to horses, and thus to transport mountain howitzers, light artillery, stores, and other military material into the heart of the mountains, would strike with a salutary terror the Comanches, Lipans, and other savage tribes upon our borders.

The habits of these Indians much resemble those of the nomade Arabs, and the introduction of the camel among them would modify their modes of life as much as the use of the horse has done. For a time, indeed, the possession of this animal would only increase their powers of mischief; but it might in the long run prove the means of raising them to that state of semi-civilized life of which alone their native wastes seem susceptible. The products of the camel, with wool, skin, and flesh, would prove of inestimable value to these tribes, which otherwise are likely soon to perish with the buffalo and other large game animals; and the profit of transportation across our inland desert might have the same effect in reclaiming these barbarians which it has had upon the Arabs of the Sinaitic peninsula.

Among the advantages of the camel for military purposes, may be mentioned the economy of his original cost, as compared with the horse or mule, when once introduced and fairly domesticated; the simplicity and cheapness of his saddle and other furniture, which every soldier can manufacture for himself; the exemption from the trouble and expense of providing for his sustenance, driving, sheltering, or shoeing him; his great docility, his general freedom from disease, his longevity, the magnitude of his burden, and the great celerity of his movements, his extraordinary fearlessness, the safety of his rider, whether from falls or the viciousness of the animal, the economical value of his flesh, and the applicability to many military purposes of his hair and his skin, the resources which in extreme cases the milk might furnish, and finally his great powers of abstinence from both food and drink. I may add another advantage, which will be appreciated by all who know the difficulty of conducting a caravan of mules or horses across the plains. I mean the security from stampedes and other nocturnal alarms and losses. The dromedary is a much less timid animal than the horse or mule, and he is not sufficiently gregarious in his habits to be readily influenced by a panic terror. The mode by which he is confined at night furnishes a complete security against escapes from fright or other causes. As he lies down, he folds the forelegs under the body. The Arab passes a loop around one or both of the folded limbs, above the knee, and secures the end of the cord around the neck. When both legs are thus shackled, the camel can rise only to the knee: if one only is hobbled, he rises with difficulty, and moves very slowly; and if an Indian were to cut the loop, and thus free the animal, and even succeed in mounting him he would not be able, without a previous practice, which he has not the means of acquiring, to put him up to such a speed as to elude pursuit. There is another point which I have never heard insisted on, but which has often struck me with some force in riding the camel. I mean the greater range of vision which, in a level country, the greater elevation of the seat gives the rider. The eye of a horseman is upon an average scarcely eight feet above the ground. Upon the dromedary it is two feet higher, and commands a wider range accordingly.

To all these advantages I am aware of no drawbacks but the expense of introducing the animal and experimenting with him, and the difficulty of accustoming horses to the sight of him. The first objection is too trifling to be debated in a case of so much importance; and though the latter has been found formidable in Tuscany, and according to Father Huc, even in Tartary, where the camel has been very long in use, yet it is of no great force as applied to the sparsely populated regions of the Far West, and as the multiplication of the animal would be gradual and slow, it is not likely that any great or general evil would flow from this source.

The facts I have recited seem to me abundantly to warrant the conclusion that at the least the experiment is worth trying, and it is highly desirable that it should be tested on a scale large enough and varied enough to embrace all the chances of success. . . .

Oration before the New Hampshire State Agricultural Society

[Delivered October 10, 1856; published in the *New Hampshire State Agricultural Society Transactions*, 1856, pp. 35–89]

The great contribution of Marsh's writing, well exemplified in this speech delivered at a meeting of the New Hampshire State Agricultural Society, was his cross-cultural perspective on agriculture and forestry. His reading habits and his extensive travels in Europe and the Middle East while U.S. Minister to Turkey gave him unique insights on both the goals and practices with respect to agriculture—"the most important of material occupations, the most indispensable of human arts"—in countries other than the United States. Marsh was firmly aware of the different time scales associated with human occupancy on different continents. Although he was unaware of the extent to which landscapes in the New World were modified by Native American civilizations (e.g., Cherokee, Aztec, Inca) prior to the arrival of Europeans, he was correct in interpreting major differences in geography between the northeastern United States and European nations as reflections of the length of time large-scale agriculture had been practiced in those two areas. Marsh's great love for the United States led him to want to help his young nation, which he felt had "hasty, impatient, and unstable habits," to improve itself and increase its standards through learning from the longer experience of others. In this, Marsh foreshadowed the animating philosophies of many conservationists in the twentieth century, especially Wes Jackson and Wendell Berry.

Marsh believed that a number of elements associated with agriculture combine to influence long-term sustainability: patterns of land ownership, crop diversification, strength of rural economies, and soil quality as related to maintenance of forest cover. He correctly understood, therefore, that agriculture could be successfully practiced only by considering a suite of social and ecological factors that interact with each other, not by considering only yield per unit area over a short span of time. Marsh's emphasis on modeling the successful practices of European agriculture indicate that his perspectives on conservation were oriented less toward protection and restoration of more-natural landscapes, which he knew had largely disappeared in Europe, and more toward rural agricultural systems that could sustainably support both crop production and communities of people with strong moral character. Although his lack of concern for biological diversity in any except the most utilitarian ways is at odds with current conservation thinking, at least he clearly defined for himself his conservation goals, a step that vexes many conservation biologists today.

The aphorism "all flesh is grass," besides its moral significance, involves or rather expresses a great physiological fact. The substantive *grass* is allied to the verb *to grow*, which, in its radical and primitive form, is restricted to *vegetable* increase and development. Etymologically, therefore, whatever vegetates, whether the microscopic mould, which gathers upon the surface, penetrates the pores and lines the cavities of larger vegetable products, the moss which tapestries the rock or festoons the wood, the herb which enamels the pastures and the woods, the cereal grain, the edible bulb, the fruit-bearing shrub or tree, and the gigantic stem of the forest vegetation, all alike is grass. Grass, or the vegetable kingdom, is directly or indirectly the sole source of animal nutrition, the only medium whereby inorganic substances are made subservient to the cravings of purely animal nature. Every movement of a limb, every breath, every pulsation, every action of every organ, every sensation, emotion or volition, every vital manifestation in short, detaches some atoms of the animal frame from their organic combinations, emancipates them from the mysterious influences of life, and brings them under the direct laws of naked chemical affinity. Each of these acts, therefore, is accompanied with an actual loss of matter to the animal which does or suffers it, and there is a constant waste of substance, which must be supplied from external sources. Inorganic nature furnishes no such supply, and vegetable processes must separate her elements, re-combine them, assimilate them, and convert them into the materials of which the animal tissues are formed, or in other words, vivify them, before they can be made to contribute to human or brute nutrition. All animals, from the invisible infusorials to the bulkiest inhabitants of air, earth, or sea, feed alike on plants or on other animated creatures, which have drawn their stock mediately or immediately from the "green herb," which in the beginning was "given them for meat," so that every creature, that lives and breathes, and "moveth upon the earth," depends at last for its sustenance and increase on that lower form of organized existence, which lives indeed, but has neither breath nor the power of locomotion. All purely physical life in fact originates in vegetation. Vegetation alone is capable of conferring the vital principle and superinducing it upon the other properties of dead matter, and we derive exclusively from plants the element that makes our flesh to differ from the "dust," whereof our first progenitor, unlike his descendants, was directly formed, and from the ashes into which the flesh is at last to be resolved.

Nor is it true, that the *life* thus given ceases with the power of assimilation, growth, and development, with the higher functions of sensation and organic movement, or with the immaterial gifts of earthly consciousness and volition. So long as the organic forces bind matter together in combinations which war with and successfully resist the laws of chemical affinity, so long that matter is instinct with life, so long it may nourish, build up and become connatural with new and successive organisms. Vitality survives sensation and consciousness and irritability, and is not extinct until chemical decomposition has resolved its products into those rudimental elements, whereof not that which lives alone, but the solid rock, the fluid waters, and the all-encompassing atmosphere also, are composed. The cloths that form the cerements of the Pharaohs, the wooden coffins that enshrine them, the papyrus that records their history, nay, the fleshly tenements of the monarchs themselves, which the art of the embalmer has made to defy the crumbling touch of time, the horned Apis they worshipped, the mammoth and rhinoceros which arctic icecliffs have preserved through whole geological periods to feed the polar bear of our own era, all these "still live," and may yet pass into and continue to live in and with a thousand successive animated forms, before the mysterious power that grouped their substance into organic combinations shall be exhausted.

Hence, AGRICULTURE, which directs and stimulates the *growth* that not only constitutes the great ultimate storehouse of all nutriment, but is the source of all organic existence, the cause and condition of all physical life, is the most important of material occupations, the most indispensable of human arts.

The estimation in which agriculture is held is a good test of the advancement of a people in civilization and the liberal arts. When Rome was at her utmost height of power and glory, her most gifted sons did not disdain to study the theory of rural husbandry, and even to give practical rules for the conduct of its minutest details. When Rome relapsed into that state of semi-barbarism, which is so apt to follow an age of great military exploit, agriculture was despised as a plebeian occupation, the laws of nature on which its successful practice rests were forgotten, and it became as unintelligent and unproductive a calling, as it was thought vulgar and humble.

In our own time, it has advanced in all civilized countries, in proportion to the intelligence of the people; and now that able philosophers have undertaken the investigation of its principles, it has again assumed almost the dignity of a proper science.

I cannot claim the merit of having added anything to the general stock of knowledge on this branch of universal industry, or even of having familiarized myself with the results arrived at by more systematic inquirers, and I can only present you with such scattered gleanings, as mere general, unscientific observation has enabled me to gather in a field somewhat wider than has fallen under the personal notice of most of you.

The remarks I have to offer you will assume the shape rather of a rambling discourse than of a formal dissertation. Their topic, so far as any special subject is adhered to, will be the general physical aspect of the most highly cultivated and densely populated parts of central and southern Europe, their agriculture and rural economy, with suggestions of improvements worth adopting, or at least testing, by us.

England will not occupy a conspicuous place in my sketches, because from the general similarity of her soil, climate, natural productions, political, civil, and religious institutions to those of the Free States of the American Union, and our community of origin and character, there is a corresponding similarity in our agriculture, domestic architecture and rural economy. Both the objects and the processes of agricultural labor are substantially the same, Indian corn being almost the only important crop not common to English husbandry and our own.

The agriculture of the two countries is differenced in *degree*, not in *kind*. There, the farms are larger, the fields broader and better drained, the tillage more thorough, the crops cleaner and more abundant, the habitations and adjacent grounds neater, the roads better graded and more smoothly kept, the horses and cattle in finer condition and more highly bred, all showing the presence of larger capital, and a constitutional, or at least habitual, tendency in the land-holder to look more to ultimate results, and less towards immediate returns, than with us.

Doubtless in all this we may find much to imitate, but since, as I have already said, not only the objects but the processes of agriculture are substantially the same as in our northern and middle States, our curiosity is less powerfully stimulated than in countries where the climate, the soil, the crops, the modes of tillage, and all the habits of rural life, are more diverse from our own, and we are less likely to be impressed with advantages resulting merely from increased care and fidelity in familiar operations, than with those which flow from novel methods, and the pursuit of new branches of husbandry.

On the European continent, on the contrary, we enter at once on a climate, a soil, a class of industrial pursuits quite different from those with which American experience has made us conversant. We encounter a people whose tastes, habits, wants, character and institutions are strange to us; whose standard of prosperity and physical and social enjoyment is other than our own, and who consequentially employ quite a different set of means for attaining the great objects of material life. On all sides we are struck by the force of contrast. The eye everywhere encounters new vegetables, and new methods of growing, securing and employing familiar plants; finds increased value ascribed to products with us deemed unimportant or found unprofitable; novel, and for the most part greatly inferior agricultural implements; habitations suited to different atmospheric conditions, and planned with a view to secure enjoyments, or to guard against inconveniences and dangers, unknown or disregarded by us.

Whatever there is of good or bad in all this novelty, whatever seems deserving of imitation or worthy to be shunned, impresses you much more powerfully, and you are more likely to derive instruction from such observation, than from viewing the nicer agricultural processes of England, which, though in general, really better suited to our condition and wants, and therefore more worthy of imitation, are nevertheless, too little contrasted with our own to excite an interest powerful enough to rouse us to a faithful study of their advantages, and which seem, besides, to demand too great a length of time, or too large an amount of perseverance and of capital, to be reconciled with the hasty, impatient, and unstable habits, and slender pecuniary means of American agriculturists.

It is a maxim, old as the time of the Roman agricultural writers, that "good farming does not pay." If by "good farming," is meant that system of husbandry which, at whatever cost, brings land into the best possible condition, and keeps it there, aims to grow the largest crops, and to raise the most highly bred stock, the maxim is true with respect to the economical results to the farmer himself, in all countries where a very dense population has not reduced labor to its minimum, and carried produce to its maximum limit of price. But it is by no means certain that such a system is, under other circumstances, equally disadvantageous to the economy of the State. In all governments where the people, the laborer included, is recognized as having rights to exercise, and interests to foster and protect, in the administration of public affairs, the power and riches of the State must be acknowledged to consist in the prosperity and wealth of its individual members, and therefore its welfare is best promoted by that public and private policy which tends to distribute and equalize, rather than to accumulate pecuniary capital. A million of dollars divided among five hundred citizens, is far more available for all legitimate governmental uses than if it were hoarded by an individual; and what is of much more importance, these five hundred independent freemen are very much more efficient and reliable supporters and defenders of the government of their choice, than if they were reduced to the condition of hirelings by the absorption of their united wealth into the coffers of a single capitalist. The riches of a free State and the riches of its people are convertible terms, and a wide, permanent national domain, or a great accumulation of public treasure, or even of invested capital, are at once usually unproductive, and at the same time repugnant to the genius of popular institutions. Those can only flourish where all public interests constitute, literally, a *common wealth*, or stock in which every citizen has both a proprietary right, and a beneficial enjoyment of control and usufruct. Now, the tendency of a system of agriculture which, like that of England, pushes its improvement beyond the point of greatest profit to the capitalist, is to the diffusion of property rather than to its accumulation, because the same quantity of land requires the labor of more hands, and thus brings a larger number to share its returns. Although, therefore, England has not the inducements to stimulate the highest amount of production, which countries like China, Holland and Belgium find in an exceeding density of population, yet the necessity of diffusing and disseminating her overgrown wealth, furnishes a reason almost as powerful for carrying the refinements of agriculture to a pitch where it ceases to be very highly remunerative to the land owner. *We* are not yet arrived at this condition. Sound economy, public or private, does not require or indeed permit us to raise the largest possible crops, or breed the highest blooded stock; and, therefore, in seeking agricultural instruction in foreign lands it is, ordinarily, the general principles, not the most highly perfected methods, that we shall find worthy of adoption.

But I proceed to my sketches. In travelling on the European continent, the first point which will strike an eye trained to geographical as well as agricultural observation, will probably be the exceeding smoothness of surface of the cultivated land. Not that the fields are flat or the inclinations regularly graded; on the contrary, the view is diversified with rocky ledge, and plain, and valley, and swelling knoll and slanting hill-side; but a long course of cultivation has obliterated the minor irregularities and inequalities of the natural surface, reduced the sharpness of the angles, removed the smaller rocks, filled the dried up water courses, and thus given the whole landscape a rolling outline, whose graceful curves, moreover, are seldom broken by hedge, or fence, or other artificial enclosure.

The next novel feature which will attract the notice of a traveller over the great and most frequented roads of Europe will be the absence of any thing which corresponds with an American's idea of a forest. In England, and in many countries on the continent, any considerable extent of unimproved and unenclosed ground, though bare of trees or even arborescent shrubs, is called a forest, the name having been retained after the proper original characteristic of the locality had disappeared. European geographical descriptions, therefore, by the different use of the term forest, often convey to American readers a very mistaken idea of the countries to which they relate. Extensive woods, partly of artificial plantation, do indeed exist on the plains of south-western France, in northern Tuscany, where endemic pestilence has wrested the soil from the dominion of man, upon the mountain ranges of both Italy and France, and in many parts of the German States; but these regions are not often visited by tourists, nor indeed do they present much of special novelty or interest, in connection with the rural economy of those countries.

Upon the principal highways of central and southern Europe, forests, in our sense of the word, are nowhere seen; and the want of them is but imperfectly supplied by the long closely planted rows of trees which fringe the road-side, and stretch along the paths and watercourses. These trees are annually or biennially, according to the luxuriance of their growth, trimmed of their lateral branches, and sometimes lopped at the top, and the chief supply of fuel, scanty at best, is derived from these poor clippings, which reduce the tallest trees to the similitude of a hop-pole, and deprive them of all family likeness to the umbrageous oaks and elms and beeches that form so fine a feature in English as well as in our American landscape scenery.

These characteristics of smoothness of surface and absence of woods are common to the most frequently visited provinces of all the countries I propose to notice, but in other particulars they are very widely diversified, whether as respects their physical geography, or the modes, habits, and objects of rural industry. To illustrate these differences I must enter into somewhat of local detail.

As France is the continental country usually first seen by American tourists, I will suppose you to land on the shores of that empire, and then to visit the Italian States, and afterwards the territories of the German Powers and of Switzerland, and I will endeavor to give you some general notion of the husbandry of each country, with an occasional illustration from other lands, and a hint or two even from distant Turkey.

After the conspicuous features I have already mentioned, the traveller in France will probably first be struck with the great minuteness of the division of the soil, and the particolored appearance thereby given to the general surface. On every side are strips of arable land, often not more than ten or twenty yards in width, and of considerable length, breaking the monotony of the plains and chequering the sunny flanks of the hills that border them. These small parcels, which exhibit every variety of agricultural product, here a miniature wheat-field or a dwarf meadow, there a bed of onions, and a plot of barley or flax, or a patch of potatoes or cabbage, the farm and the garden intermingled, belong to different proprietors or at least occupants, living often miles from their petty fields, and of course cultivating them at no trifling disadvantage. The modern law of descent in France which excludes primogeniture from sole inheritance of the soil and divides the estate among all the heirs, combined with the tenacious attachment of the French peasant to his country and his paternal acres, has occasioned this exceeding minuteness of partition, which has been found productive of important political advantages, and at the same time of serious economical evils, both of which are curiously contrasted with the opposite benefits and inconveniences resulting from the monopolizing of lands in England by the operation of the British laws of inheritance and entail.

The farmer whose estate embraces but an acre cannot keep a flock of sheep, or a yoke of oxen, or scarcely a pig; he cannot build barns or sheds for storing his crops; he cannot afford a comfortable dwelling, or possess a cart, a wagon, plows, harrows or even a full set of the smaller agricultural implements. He cannot spare ground for the growth of fuel, underdrain his land, or secure it by a permanent or substantial fence, and he has too small an interest in internal improvements to make him an able or a willing contributor to roads and other facilities of travel and transport. He must live, therefore, in such a hovel as his slender means enable him to own or rent, at so great a distance perhaps from his narrow territory, that half his day must be spent in going and returning between his dwelling and his place of labor; his ground must be painfully tilled with spade and mattock; the extra aid he may require at critical seasons must depend on the uncertain chances of an exchange of labor; the necessities of his household or the urgency of the tax-gatherer may compel him to sacrifice his own harvest, while earning a hard penny in securing that of his wealthier neighbor; and whatever he gathers at last must be sold on the spot at an inadequate price, or transported to market or to his garner upon his own shoulders or those of his wife and children.

In such districts, therefore, you see no flocks of sheep or herds of cattle, but here and there, nibbling by the road-side or along the *lands*, two or three or perhaps a dozen ewes kept from straying by dog and shepherd, or a single cow with a rope about her horns held by a woman who divides her time between tending an infant or knitting a stocking, and checking the obstinate propensity of her beast to yield to the temptations of cabbage and clover and trespass upon that which is her neighbor's.

At distant intervals you pass hamlets or clusters of little huts with thatched and moss-grown roofs, thickly planted on narrow and crooked streets, with a picturesque old gothic church, and a few venerable trees, the benefits of whose shade have saved their spreading boughs from the process of lopping, to which their fellows are elsewhere subjected. The rural population was originally collected in these villages for security during the troublous times of that fearful period of spiritual and temporal tyranny, the middle ages. They gathered in huts nestling under the shadow of some rocky hill crowned with a monastery or a feudal castle now generally destroyed, or existing only in the mouldering fragments of its donjon-keep, which was too solidly built to yield even to the violence of the French Revolution. In spite of the inconvenience of so distant a separation between landholder and land, these villages continue to exist, partly from habit, partly from the national passion for social enjoyments, and partly from the inability of the villagers to provide for themselves new and more commodiously situated habitations. Occasionally, indeed, you pass a new chateau or an old castle modernized, surrounded with shady groves and flowery gardens, orchards of stiff pear trees or ungraceful prunes, with flocks and herds, teams of horses and oxen, and all the larger and more costly implements of tillage. But these are rare, and it is but seldom that even they are accompanied with neat and comfortable rural homes for steward and laborers or other dependents.

The average quantity of ground belonging to the landed proprietors of France is not above five acres to each, and as very many estates greatly exceed this average, very many must fall much below it. Although, as I have already shown, with such small proprietorships the best modes of farming cannot be adopted, yet from the variety of crops, contemporaneous or successive, that may be obtained from lands cultivated in the fashion of the garden rather than of the farm, and with the spade more frequently than with the plow, from the care taken to make every inch available, and above all from the fact that the eye and the arm of the owner of the soil are both employed in bringing out its utmost resources, the actual quantity of produce applicable to human sustenance is great, even as compared with that of the best farmed land in Europe. In this, as in many other instances, we are in danger of being misled by statistical tables, which generally notice only marketable products, whereas these small French fields are made to yield a variety of plants abundantly useful in domestic economy, but either not employed by the citizens of large towns, or, in each individual case, reared in too small quantities to find their way to market, or to attract the notice of the political economist.

The most valuable results of this system, however, are not economical, but moral and political. It is a matter of immense consequence to the State, that they, who in the first instance extract from the bosom of the soil that which supplies the life-blood of the nation, and who in the last resort are the defenders of that soil, should feel that they are something more than tenants at sufferance upon it, and that they have a substantial proprietary interest in the earth which nourishes them, and through their labors those classes of their fellow citizens who "toil not, neither do they spin."

Since the Revolution which broke up the great feudal estates, and divided them amongst the peasants, who before had been but serfs upon them, there has been an immense change in the character and condition of the French husbandman. The rural laborers, in spite of the liberal and philanthropic views of Henry IV, whose dearest wish it was that every peasant might be able to have a fowl for his Sunday's dinner, were degraded far lower than they ever were in freer England. "We see," said La Bruyere two hundred years ago, "dispersed over the country certain savage animals, male and female, with dark and livid skins, naked, scorched by the sun, bound to the soil which they dig and stir with obstinate perseverance. They have a sort of articulate voice, and when they raise themselves on their feet they show a human face; and in fact they are men. At night they retire to dens, where they subsist on black bread and roots. They save other men the labor of ploughing and sowing and harvesting, and therefore are entitled to some share of the bread they have sown." A later writer (Courier) remarks that in this extract La Bruyere was speaking of the more fortunate class of peasants, those namely who were blessed with bread though black, and the opportunity to earn it, and these, he adds, were the fewest.

Such, making some allowance for rhetorical exaggeration, were the farmers of France, at the period when your enlightened ancestors were laying the first foundations of our mighty empire on the Atlantic coast of New England. Now, however, the tillers of French soil, as in the northern United States, know that they constitute the weightiest element in the State, and they are hourly making themselves more and more felt in the political action of their government. This consciousness of the importance and the responsibilities of their position, and their consequent identification with the power and glory of France, is one of the main reasons why so few Frenchmen emigrate to the New World, and it serves to explain how in the late destructive war, there was comparatively little difficulty in recruiting the French army with men whose rural life had given them the best physical training, and the proper moral qualities to form a spirited and efficient soldiery.

In Great Britain, on the contrary, the accumulation of the real estate of the island in the hands of a comparatively small number of proprietors, has produced effects quite opposite to those which I have just described as existing in France. The causes of this accumulation I cannot here discuss in detail, but I may mention in general that they may be found, as respects northern Scotland, in the abolition of the patriarchal system of the Highlands, by which the chief was considered as holding his lands in trust for the clan; and elsewhere, in the operation of the law of primogeniture and entail, the great increase of capital from trade and manufactures; which has sought investments in lands, and the extension of pastoral husbandry, sheep breeding especially, which has converted millions of acres of well peopled arable soil into broad pastures, inhabited only by the shepherd and his flock.

The estates of the great proprietors are doubtless more scientifically cultivated than when they were divided into smaller portions, and the liberal application of capital and intelligence has carried English agriculture to a pitch of perfection perhaps never on a great scale realized elsewhere. The marketable products, and probably the total material which contributes to human sustenance, have been largely increased, but yet, while the entire population has doubled in fifty years, the rural districts have in many instances lost almost all their inhabitants. Hundreds of thousands of the classes whom it was important to retain, have migrated to America, Africa, Australia, and New Zealand because they could find no rest for the sole of their feet on the soil that bore them. Others are absorbed by the demands of the manufacturing system; others in the construction and working of railroads and their accessories; others in the increase of the commercial marine, and others, again, enter into that vast accretion, for it cannot be called organic growth, by which the great cities of England have swollen to such unnatural and disproportionate bulk.

Thus the class of small proprietors and tenants upon long leases, who cultivated their own lands, and who were justly regarded as the best reliance of the State in every critical emergency, has now quite disappeared, or been greatly reduced in numbers, in moral force and in physical power. With the smaller agriculturists have vanished also the small tradesmen and mechanics. Not only are homespun clothing and all household manufactures obsolete, but every rural implement, every article of clothing, from felt to sole-leather, is brought ready-made from some great town, and the village blacksmith and tailor exist no longer. The inconveniences of this state of things had not been distinctly felt and appreciated in England until the fruitless efforts to raise men for military service in the East showed that the English, in ceasing to be a rural, had ceased also to be a martial nation, and that a manufacturing and a civic people have neither the moral nor the physical qualities which made the British armies so formidable in their wars with Napoleon.

Whether these evils are compensated by the advantages resulting from division of labor, augmented mechanical power, and other means of increasing the total industrial product of the empire, it is perhaps too early to pronounce. But it must be remembered that in all history, the growth of cities at the expense of the rural districts, which is becoming noticeable even in our young republic, has proved a token and a near precursor of national decrepitude and decay. The city-bred youth, with an enfeebled physical frame, has a precocious development, a one-sided and premature culture, a quickness of intellect, a promptness of movement, and an unnaturally stimulated sensibility, or rather sensitiveness, which give him an apparent superiority over his calmer, slower, more impassive bother of the country. But this superiority is at best a superficial and unreal advantage. Man's true strength of mind and body, his physical, moral and intellectual nature, are most completely, equally, and harmoniously evolved, trained and perfected in modes of life akin to that which God prescribed to our first parents when he made them tillers of the ground.

. . . The fields in [*Western*] Italy are irrigated by canals derived from streams whose beds not unfrequently lie above the level of the plains they traverse, and which are kept from overflowing the champaign and converting it into pestilential marshes only by high and costly embankments. You are now struck with the general characteristic of the physical geography of Western Italy, which distinguishes it so remarkably from most other mountainous countries. The surface is not composed of hills and dales with level plateaus on the ridges and narrow threads of what in America we call intervale bordering the water-courses at the bottom of the valleys between, but it consists of wide spread plains and abrupt mountain elevations, with no intervening gradual rise to break the suddenness of the transition. The plains are generally at but a small elevation above the sea, and though in some instances probably of submarine formation, yet in many others they appear to owe their origin to the action of the torrents, which, after the Apennines were bared of their forests, gradually washed down the vegetable soil and disintegrated or decomposed rock, first into their own estuaries, and then, when these were filled to the water level, deposited it on a broader surface, until a wide and continuous belt of champaign country was interposed between the mountains and the sea.

Eastern Tuscany, being more distant from the sea and more elevated, is less strongly marked by the features I have described; but the tendency to form extensive levels and sudden ascents still characterizes its geography as it does that of the Pontifical states west of the Apennines. The climate of the territory of the Church is somewhat milder than that of Tuscany, and the funeral cypress, the myrtle and the laurel, all evergreens, now become frequent and conspicuous ornaments of the pleasure grounds and gardens. Some miles before reaching Rome, you enter the famous Campagna di Roma, a fertile but pestilential plain, now almost destitute of trees and rural habitations, in which the Eternal City is embosomed. The Campagna is, to a considerable extent, of igneous origin, but to the south and west it gives place to the Pontine marshes, which are composed of marine or alluvial deposits. The whole of these extensive regions, the former of which has a certain resemblance to the rolling, the latter to the flat and more swampy prairies of the West, would admit of unlimited agricultural improvement, and again sustain, as they once did a dense population, were it not for the fevers which render them almost uninhabitable, and are threatening to depopulate the city of Rome itself. For this evil, science has failed to find a remedy, as it has been unable to detect its cause; and though considerable winter crops are grown upon the Campagna, yet the greater portion of it is occupied for the pasturage of herds of half-wild oxen, sheep, and horses, which are driven to the mountains for pasturage in summer, and brought back on the approach of winter. The buffalo alone passes the hot as well as the cold season in the Campagna, and, with the exception of the mountaineers who venture down to gather the crops in harvest, at the imminent risk of their lives, and a few pallid wretches, who remain to watch the property of their employers through the summer and autumn, or to attend at the government post-stations, the whole of this vast tract is quite uninhabitable during the half of the entire year.

Leaving the Campagna and the Pontine marshes, you traverse a more diversified region and soon enter upon the fertile plains of Neapolitan Campania or the Terra di Lavoro, which are almost wholly composed of ejections from the volcanic group whose only present active crater is that of Vesuvius. The soil of this province is exuberantly productive, and its climate is more salubrious than that of the Roman Campagna. The vine, which, in more northern Italy, had, as we have seen, already begun to assume more freedom and luxuriance of growth, is here allowed to climb tall trees planted in rows for that purpose, and the shoots from neighboring trees are interlaced so as to form a net-work or vast arbor high over head. The ground is cultivated with a variety of plants, and such is the fertility of the soil and the power of the sun, that good crops of grain are raised in the vineyards under a depth of shade, which, in less favored climates, would admit the growth of no vegetation but spontaneous underwood.

The immediate environs of Vesuvius, where the soil is of unmixed and recent eruptive origin, possess the usual fertility of volcanic earth, and the lavas and scoriæ of this mountain seem in general to yield to atmospheric influences and to become sufficiently disintegrated to admit of cultivation sooner than those of Etna. The Etnæan lava of 1669, as well as the beds of volcanic sand and ashes deposited near the outlet from which that terrible current issued forth, seem almost as bare and as black as on the day of the eruption, while ejections from Vesuvius, of a much later date if we can trust the report of the inhabitants, are already covered with vegetation. The soil upon the flanks of these mountains presents a character to which we have, at least on this side of the Rocky Mountains, nothing analogous, as all our igneous rocks and the earths formed by their decomposition are of an earlier origin and a different constitution.

The configuration of the surface in that portion of central Italy which lies east of the Apennines presents more points of analogy with our own geography. There is less of widely extended plain than upon the western shores of the peninsula, the mountains rise more gradually from the sea, and the landscape is chequered with hill and knoll and winding rivulet and narrow intervale, but the objects and processes of rural industry are much the same as on the opposite slope of the mountains.

Proceeding northwards we enter the vast alluvial plains of Romagna and Lombardy, which are watered by the Po and the Brenta, with other rivers and their tributaries, and extend hundreds of miles in both directions. Neither the soil nor the climate of this region is so well adapted to the grape as those of central and southern Italy, but the vine still flourishes on the flanks of the mountains that border the great plain, and the few detached elevations that break the general uniformity of its surface. The orange disappears altogether. The mulberry becomes more and more important. The pasture grounds and the meadows are of unsurpassed productiveness. Indian corn is very largely grown, and preserved through the winter by being *tressed up* on the ear, as seed corn is with us, and hung out in the open air under the projecting roofs of the dwellings and out-houses. Mush or hasty pudding, under the name of *polenta,* here plays a very conspicuous part in the nutrition of the laboring population. Rice, too is produced in abundance wherever the soil can be flooded at pleasure, and this grain thrives even as high as 45°, but its cultivation is so prejudicial to the health of the country, that government has found it necessary to forbid its extension. The beds of many of the principal streams are considerably elevated above the plains through which they flow, and thus furnish a convenient supply of water for irrigation, for navigable canals, and for mechanical purposes. The complicated net work of canals, by which the ground is drained in wet seasons and the crops irrigated in the dry, is believed to date earlier than the Roman conquest, and practical hydraulics, or the art of directing and controlling the flow of running water is nowhere better understood than by the engineers of Lombardy. The plains of Lombardy and Romagna appear to have been formed almost wholly by the deposits of the Po and other rivers whose affluents rise in the Alps and the Apennines, and they are now in a course of very rapid extension from the operation of the same cause. The encroachments of the land upon the Adriatic are obvious to the most careless observer, and it may be cited as an instance of the rapidity of this process, that the town of Adria, which was a seaport 200 years after the time of Christ, is now fifteen miles from the water.

. . . A very striking and important general difference between the agriculture of most of these countries and that of the United States is, that while all our crops of any commercial value with the exception of the grasses, are the product of annual vegetables, a very large proportion of the profits of European continental agriculture is derived from perennial plants. Thus the vine is common to some of the districts of all the countries I have mentioned; the mulberry and the olive to most of them. In France the annual yield of wine and brandy is estimated to be worth more than sixty millions of dollars, and that of manufactured silk something above forty millions, while that of the cereal grains does not much exceed two hundred millions. If to the products of the vine and the mulberry we add those of the olive and the cork-oak, we shall find that these perennials, all of which well might be, but none of which yet are, cultivated in the United States to such an extent as to be of any national importance, yield in France returns much more than half as great as those of the grain harvest of that empire, and of considerably greater value than the entire unelaborated cotton crop of the American Union.

It is obvious that this difference in the objects of rural husbandry must make an essential difference in the character of agricultural labor and the occupations of the people. Not only are the toils of those employed in cultivating perennial plants lighter in themselves than those of the husbandman whose seed time as well as his gathering is annual, but the mildness of climate, which renders the cultivation of perennials practicable, facilitates the labors of agriculture, by allowing them to be continued through the year, instead of being crowded, as they are with us, into the compass of little more than a season, and at the same time, it increases the rewards of industry by admitting a succession of crops which makes the whole year a perpetual harvest. In the same countries the absence or rarity of frost both exempts the people from the expensive and laborious necessity of providing a large supply of fuel, and permits the growth of pasture-grasses through the winter, whereby the cost and toil of securing and feeding out a stock of hay or other winter fodder is in a great measure avoided. Other circumstances which tend to lessen both the labors and the anxieties of rural life in such climates are the slowness of vegetation, and the general dryness of summer and harvest time. Where the changes in the condition of the crops are very gradual and the weather almost certainly fine, agriculture has few critical periods, and those occasions so common with us, where an unavoidable delay of a day or two involves or hazards the sacrifice of a crop, are of rare occurrence in southern Europe.

In provinces devoted especially to the growth of the mulberry and the rearing of the silk-worm, there is an exception to this remark founded however not so much on the uncertainty of the weather, as on the necessities and the habits of the animal. Through the feeding season which lasts from about the middle of April to the end of June, the whole rural population, old and young, is absorbed in this single occupation, which requires little outlay of physical strength, but makes very large demands on the intelligence and watchfulness of the laborers. "During this period," says a French writer, "all other labors cease. We neither buy nor sell. Legal proceedings are suspended. Everything is adjourned which can possibly be postponed. Merchants, notaries, lawyers, doctors and apothecaries, all take their holiday. The peasants have not even the time to be sick."

The vintage, too, is a season of great activity, but its toils are light, and it is always regarded more as a joyous and festive occasion than as a period of unusual care and labor.

Farm lands are very seldom enclosed, both cattle and sheep being always watched and prevented from straying and trespassing by herdsmen and shepherds, usually aided by dogs, and when the flocks pasture far from the dwelling of the proprietor, accommodated with small lodging houses, which are moved to and fro like a barrow or a hand-cart on a pair of wheels. Thus the capital and labor which we expend in building walls and fences are almost wholly saved.

Besides all this, the great permanent improvements, the clearing of the forests, the drainage of the soil, the smoothing of the ruggedness of its natural surface, the building of terraces, the planting of the perennial vegetables, the construction of houses and churches and roads and bridges, all these have been substantially accomplished by former generations, and the agriculturist has none but the easy labors which the changes of the seasons impose. Remove the curse of temporal and spiritual tyranny and misgovernment from climes like these, and the poet might indeed well exclaim,

> "Ah, happy husbandman, didst thou but know
> "The blessings of thy lot!"

Some of the most important branches of rural industry in southern Europe, as well as in other regions, are threatened with serious damage from sources altogether new, or which, if known in former ages, have passed away without leaving a record behind them. The grape disease, like that of the potato, long baffled all attempts to check its progress, and though much mitigated in some provinces, it still menaces the entire destruction of the vineyards in many of the best wine growing districts in the world, and, consequently, great changes in the habitual occupations of millions. Originating in an English hot-house about the year 1845, it has spread over the whole of southern Europe, including the Ionian islands and Greece, where it has almost wholly destroyed the crop of the small seedless grape known in commerce and confectionery as the Zante currant, and it is now advancing eastwards and northwards, producing the ruin and too often the utter destitution and starvation of thousands who have no resource but the vineyard. Independently of this new disease, there had been before a suspicion that the grape, as well as many other domestic plants, was degenerating and destined to final and not distant extinction. It has been said that the clusters are smaller and less abundant than in earlier times, and that vines planted within the present century have not grown with the ancient luxuriance. There does not appear to be any sufficient evidence in support of this belief, but it would certainly be difficult to find vines of such dimensions as are known to have formerly existed. The cathedral at Ravenna had a great door composed of planks of vine wood, thirteen feet long and fifteen inches wide, as the skeptical may be easily convinced by examining the remains of them still preserved in that building. They are said to have been brought from Constantinople many centuries since, and to have been taken from wild stocks growing on the banks of the Rion, the Phasis of the ancients, which flows into the eastern end of the Euxine. The vine still grows on the same river with greater luxuriance than in any other known locality, but hardly attains the enormous dimensions required to furnish such planks as I have described.

The olive and the orange, too, are suffering from epidemics somewhat resembling in their effects the grape-disease; but whether, like the latter, these maladies result from the growth of a microscopic parasitic vegetable, or whether they are produced by an insect, as in our Florida orange, is matter of dispute, and it is not improbable that both causes are at work in different localities. The silkworm has always been liable to many disorders, but a new disease has made its appearance in France, and excited a good deal of alarm among the silk-growers. It has no very distinct symptoms or characteristics, and appears to consist rather in a general degeneracy of the species arising, as some conjecture, from the want of crossing, or breeding *in and in,* as it is technically termed. It is, however, too recent to have been yet thoroughly investigated.

From the prevalence of these evils we may draw the important lesson, that sound policy requires a considerable range in the objects of rural industry, in order that the failure, temporary or continued, of any one branch of husbandry, may not produce the ruinous effects, which have resulted in many European countries from a too exclusive devotion to a single agricultural pursuit, and the consequent dependence of a whole population on the contingencies of one method of earning its bread.

Having thus viewed the general surface and most striking physical characteristics of several of the principal continental countries, let us look at some points of their rural economy a little more in detail, and see what is most worthy of imitation in their public improvements and their private husbandry. I cannot of course on this occasion go into minute particulars, or notice all the specialities, which one having time and adequate motive to devote himself to individual fields of agricultural and industrial inquiry would bring out, and I must content myself with selecting a few obvious points of comparison, and which seem to me interesting as evidences of our superior progress in some particulars, or on the other hand suggestive of improvement in our rural life.

A striking point of difference in the industry of Europe and the northern United States is the greater efficiency of the laboring man in the latter. This arises partly from physical causes. Among the former are the superior intelligence and ingenuity of the American laborer, and his consciousness that he is performing a voluntary contract for an adequate consideration fixed by his own agreement, and not by the pleasure of an arbitrary feudal lord, or by external circumstances which allow him no option but half-remunerated labor or starvation; among the latter, are the more nutritious diet of our farmers, and the great superiority of the tools and mechanical appliances employed in American agriculture. Thus, in many countries of Europe, the hoe has a handle but three feet in length, with a blade of half the width and thrice the weight of our own, while the axe is scarcely half as heavy as that which our wood-choppers wield with such powerful effect; the rake and pitch-fork are of the clumsiest fashion, and the latter wholly of wood; the scythe blade is but two feet and a half in length, broad and heavy, forged of soft iron, and sharpened, not by grinding or whetting, but by hammering the edge on a stone, another stone often serving as the hammer; the snath is straight and with a single handle, the left hand grasping the end of the snath; and so ignorant are the peasants of mechanical powers, particularly in southern Austria, that I have seen four men exert for a long time their utmost strength in trying to shove a stone over the rail of a wagon body on a pair of skids, when one man, by raising the lower end of one skid with each hand, would have tilted the stone into the wagon with entire facility.

Although public works of internal improvement do not strictly belong to the subject I have proposed to myself, yet they are intimately connected with the progress of agriculture; and the development which they alone can give to the physical resources of any extended territory, essentially effects every branch of rural economy. They may, therefore, appropriately be considered in connection with our more special topic, and it will not be amiss to contribute something to the correction of a widely diffused popular error. In our indiscriminate national self-esteem, we are apt to imagine that the excellence of our political institutions has extended itself to all our national undertakings, and that the builders of our canals, railroads and highways are as superior to European engineers in constructive skill as the framers of our federal constitution, to the deputies of the convention in the French revolution, in political wisdom; but this is an assumption by no means yet warranted by proof. What our engineers might do with a larger command of time and means, remains yet to be seen; but the utmost they have yet accomplished in the way of internal improvements, with the almost solitary exception of our wooden bridges, not only falls short of what has been effected upon every important railway, but finds its parallel upon almost every great common road in Europe.

The European public works are generally superior to ours, both in boldness of plan and in thoroughness and fidelity of execution. The canals, whether for navigation or for irrigation, exhibit an intimate familiarity with the laws of hydraulics; and the masonry and all the appurtenances are usually of the most finished, solid and permanent character. The railroads exhibit even a more marked superiority over our own. The line is usually admirably planned; difficult grades and long circuits are avoided by tunneling, which is carried so far that it is not uncommon to pass through eight or ten miles of tunnel in a single day's journey; the track is always double; all embankments and the scarp of all earth cuttings are either sodded or paved, thus avoiding the annoyance of dust and the danger of slips and slides; the bridges are usually of stone, and the masonry of these as well as of the viaducts, some of which, as at Venice, are more than two miles long, is of the best possible workmanship; the cars are most commodious, and the number of guards, brakemen, conductors, engineers and signal-men, is such as to give every security against disorder or danger.

But I am inclined to think that there is no single fruit of high civilization and long continued social order, which more frequently excites the admiration and surprise of an observing and intelligent American in Europe, than the condition to which the common roads have been brought, and in which they are maintained, by a great and persevering expenditure of money and of skill. Many of these were laid out by the Romans, and still, to a considerable extent, follow the lines engineered by those masters of the world, though in general the routes have been more or less altered, the more widely diffused use of wheel carriages requiring a considerable reduction of grade. The rise is seldom allowed to exceed three degrees, or one foot in nineteen, and to obtain this, long circuits, a continued series of zig-zags, or tunnels, sometimes of great length, are resorted to. The traveler sometimes travels a mile to gain a furlong; but as the roads are always wide enough to admit of passing without hindrance or delay, securely walled or fenced with masonry, well macadamized, so as to be hard and dry at all seasons, and of such moderate grades that horses can always trot rapidly down with safety, and go at a good pace up, the actual distance accomplished is as great as it would be over a straighter road with steeper grades, and with infinitely less fatigue to man and beast, less wear and tear of harness and vehicle, and greater security of life and limb. In many instances the roads are cut with great cost and labor, in the face of a cliff, and I suppose that more than half the famous road called the Cornice, between Nice and Genoa, a distance of 140 miles, is so built.

For the construction of such expensive works, European governments, beyond reasons of general convenience, have had a motive hardly operative here. I mean the expediency, not to say the necessity, of providing employment for the poor, and thus sustaining them at the expense of the rich, by a method less obnoxious than a pauper tax, and at the same time promoting the public service, and what is equally important, attaching the laboring classes to the power that gives them bread.

We have been too much absorbed in the grander and more obvious improvements by railroads, to attend to the equally important subject of bettering the condition of our common highways. Doubtless it is an advantage to be able to send your surplus from the depot to the market in one day, instead of three or four; but you who live twenty miles from the station would gain almost as much by so improving your common roads that the team which conveys your produce to the railroad, instead of spending two days in going and returning between your farm and the depot, could go and back, with double or treble the load, in one.

Sound economy dictates the policy of extended internal improvement in our Eastern States, both as a means of making our lands more profitable to ourselves, and of rendering them more valuable and desirable in the eyes of others. The attractions of the seductive West are draining us of our pecuniary savings, and of that far more valuable capital which consists in the moral, intellectual and physical energies of our ambitious and enterprising youth. To check this mischievous drain of money and of men, which not only locally impoverishes us, but weakens the nation, by diffusing its population and its capital over a wider space than they are able to occupy to the best advantage, is an object well worthy the thoughtful consideration of every patriotic New Englander, and we could well afford to make great sacrifices to accomplish it.

The superiority of European public works to our own, whether as respects the plan, the execution, or the management, is so great that one can hardly travel over them without being led seriously to doubt whether our system of leaving the construction and control of lines of internal communication to private enterprise is not an erroneous one, and whether they ought not in all cases to be undertaken and managed by the general and state governments, rather than entrusted to the hands of irresponsible and, as experience has shown, for the most part unprincipled, speculating corporations.

The effects of corporate action in these matters have been much the same in England as here; while on the continent of Europe they have been chiefly avoided, by holding all internal improvements in the hands, or at least under the complete control, of the authorities of the State.

On the other hand, it must be admitted that the execution of public works by the general or State Governments in America would be liable to one result mischievous here, but which is rather a benefit in Europe, or at least is neutralized by advantages important enough to counteract, or rather compensate, its evils. I refer to the great increase of governmental patronage, and consequently of political corruption, which belongs to their system. The management of our State works is perhaps as much infected with political depravity as any branch of our national government; but it deserves to be well considered whether even this is not a less evil than the wide-spread demoralization and the vast amount of private ruin and misery, which are necessary consequences of the predominance of corporate action and the trade of stock-jobbing. At any rate the evils of government patronage might be lessened by lengthening the term of office, and making it more independent of party favor; and if you were surrounded by a greater number of partizans ready to beguile you of your vote, you would probably find fewer whom practice had taught dexterity in the act of lawfully picking your pocket.

In continental Europe the amount of government patronage has been greatly increased, by the extension of internal improvement and of the post routes; but though the governments have thereby secured a great number of active and intelligent agents, apparently interested in their support and devoted to their maintenance, yet it must not be forgotten that these agents come from the body of the people, and are greatly influenced by the sentiments, and alive to the interests of the class from which they spring. The governments are thus brought more directly into contact with the people. The petty officers connected with the railroad and the post office form a sort of plebeian aristocracy, which serves as a link between the highest and the lowest, the governors and the governed, influencing and influenced by both ends of the chain alike; and though they are doubtless, to a considerable extent, a means of intimidating and politically corrupting the people, they on the other hand serve to instruct the rulers in the true condition and wants of the lower classes, to mitigate the asperity of feeling, with which the humble are regarded by the proud, and thus somewhat to soften the relations between arbitrary rulers and a down-trodden people.

But I am indulging in too widely discursive a strain, and will return to my more immediate subject, and proceed to notice some practices connected with rural economy in Europe, which, with such modifications as circumstances may require, seem to me perhaps worthy of imitation.

Much attention is paid in Europe, both by governments and by individual proprietors, to the renewal and preservation of the forests. Hundreds of acres are annually planted with oaks, pines, larches, and other timber trees, and Europe will be better supplied with wood in the next century than it is in this or even was in the last. In most private forests, small portions are cut regularly, at intervals of from fifteen to twenty-five years; and in situations where the clearing of the land would lead to injurious consequences, such as the washing or sliding of earth, or the fall of avalanches, it is often forbidden altogether.

I am aware that this subject has recently been much discussed in the United States, but its importance is not yet generally appreciated nor can it be, but by the careful study of the matter in countries where time has been allowed for the full effects of the destruction of the native forests to develop themselves. We are already beginning to suffer from the washing away of the vegetable soil from our steeper fields, from the drying up of the abundant springs which once watered our hill pastures, and from the increased violence of our spring and autumnal freshets, to say nothing of the less obvious meteorological effects of too extensive and injudicious clearing; but it is only in countries that have been laid bare of their natural clothing for generations, that the extent of the devastation thus produced can be comprehended.

There is no doubt that nearly the whole of the Apennines, as well as the lower slopes of the Alps and the Pyrenees, the mountains of interior Spain, of the Mediterranean islands, of Greece, of Asia Minor and of Syria, were at one period covered with timber. They were chiefly stripped of their forests in remote ages by human improvidence, and the consequences have been in a high degree disastrous. The most obvious of these has been the increased rapidity with which the rain water and melted snows are carried off, the consequently augmented violence of the torrents in the rainy season, and extensive degradation of the soil and denudation of the rock at the higher elevations. The arable land of whole provinces has thus been laid waste, and though wide and fertile plains have been formed by the deposits left by the subsiding currents, yet extensive regions have by the same cause been converted into pestilential swamps, and become entirely uninhabitable. Where the rock has been once laid bare, or the remaining earth deeply furrowed, it appears to be no longer possible to cure the evil and check future ravages; but the preservation of the forests on mountain slopes where they now grow is thought to be a secure safe-guard against the extension of the mischief. But although in some localities these devastations can no longer be prevented, the ingenuity of Italian engineers has found a means of turning them to good account, and of compelling even the mountain torrent, the very symbol of uncontrollable fury, to repair or at least compensate its own ravages.

The enterprises to which I allude are among the most remarkable triumphs of humanity over physical nature, and they possess special interest as exhibiting almost the only instance where a soil, which man has once used, abused, exhausted, and at last abandoned, has been restored to his dominion, re-occupied and again made subservient to the purposes of social and industrial life. I refer to the success of the Tuscan engineers in reclaiming a very large extent of marsh in the Val di Chiana, in Tuscany, by processes which have been since employed in other parts of that duchy, and the application of which elsewhere might save vast territories from disease, sterility and desolation. The streams which flowed through the Val di Chiana had, in consequence of the gradual elevation of their channels and the filling up of the bed of the valley with gravel and earth, overflowed their banks and transformed many square leagues of ground into a barren and unhealthy marsh. Fossombroni undertook to reclaim these swamps, and succeeded in restoring them to fertility and salubrity, by erecting dams, embankments, and sluices in some places, and cutting water-courses in others, so as to obtain the complete control of the waters of the valley, and thus compel them to deposit, at pleasure, the mud with which they were charged in the inundations of winter. By this means, the low grounds were first filled up, and then, by elevating the embankments and dams, deposits were formed at still higher points, and thus not only was the general level of the valley considerably raised, but its inclination was so changed, that the course of the streams which water it was reversed, and the Arno now receives affluents which, from time immemorial, had discharged themselves into the Tiber. The quantity of land already reclaimed by these operations was estimated, in 1835, at an extent not less than ten American townships, and this has been considerably increased in succeeding years.

. . . Let me here enter a protest against a single European abuse out of the many which might be specified; and then, after a few general remarks on the benefits which agriculture may be expected to derive from physical science and the extension of geographical knowledge, and upon the spontaneous vegetation of Europe, I will release you.

The abuse I referred to is the almost universal employment of women in field work on the continent of Europe. I have rarely seen them act as teamsters, or mow or hold the plow, but there is scarcely any other species of agricultural labor, which is not in the largest proportion performed by women. I have often seen them even carrying stone, gravel, and earth, for repairing the roads, in baskets on their heads, and in one instance, I observed the building of a very heavy railroad embankment almost exclusively by this method; about five hundred women and boys being occupied in transporting the earth for the filling, a distance of about five hundred yards up a steep ascent. Field labor is not only prejudicial to the health of women, but it tends irresistibly to deprive them of the softness and grace of their sex, to assimilate them to the coarseness of the men with whom they work, to disqualify them for the duties appropriated to them by nature, and in short to debase and brutify their whole character. It is well known that women are nowhere treated with so much consideration, deference, and respect as in the United States, and I believe their exemption from field labor, and their consequent disconnection from all the grosser and more repulsive cares and toils of husbandry, has much to do in fixing the social position they so well merit, and happily for the true interests of our own sex, so fully enjoy, throughout the United States.

Let me now indicate one or two points in which our rural economy may be profited by availing ourselves of the advancement of natural science.

The efforts of agriculturists have been hitherto mainly directed to the attaining of the greatest quantity of produce, without sufficiently inquiring whether the very means employed to stimulate extraordinary fertility did not deteriorate the quality, in nearly as great a proportion as they augmented the yield. There are some facts connected with this question which are familiar to every one. Our native simples which are gathered for medicinal purposes are much more efficient and beneficial in their action, when growing untilled on the barren soils where nature usually sows them, than in the rank and vigorous form they assume when transplanted to the too luxuriant soil of our gardens. So the pasturage and the hay crop are so much more highly flavored in dry, than in moist and fruitful, seasons, that their superior nutritiousness seems sometimes quite to compensate for their diminished quantity. In both these cases, the facts are easily tested by simple experiment; but this becomes more difficult, when we attempt a comparison between different kinds or qualities of any of the grains employed as food for man. The problem is here too complicated to be solved by ordinary observation, because, in countries where any enlightened interest is felt in such questions, men are seldom confined to any one article, or any specific and minutely ascertained quantity, of diet. For determining the nutritive properties of our aliments then, we must have recourse to chemical analysis, combined with well devised experiment. By means of analysis we learn at once that the nutritive ingredients in different specimens of the same article of food, and, of course, the actual value of the article, vary very widely. The chemical constitution of the cereal grains, wheat for example, is by no means constant, and the amount of nutrition yielded by a given quantity of this grain is modified by the meteorological character of the season, the qualities of the soil and of the fertilizers applied to it, the mode of cultivation and the time of harvesting, as well as by many obscurer causes. Thus, wheat generally contains sixteen or seventeen per cent of gluten, which is easily separated by washing; but there are varieties, or occasional crops of wheat from which no gluten can be obtained, and there are others where some of the berries from the same seed yield the usual proportion of gluten, the rest are apparently without it. If the flour is used for bread, this difference in the character of wheats is not readily noticeable, or if observed it is usually referred to other causes than a difference in chemical character; but in countries where macaroni, which can only be made from wheat abounding in gluten, is largely manufactured, the proportion of this ingredient becomes at once a matter of familiar observation, and of very serious importance. Hitherto no tests of the true value of different samples of grain, at once certain in their indications and easy of application, have been discovered, and wheat and other cereals are at present judged of in the market only by their external characteristics. It seems not improbable that, with the advance of organic chemistry, some ready means will be devised of determining the proportion of nutritive ingredients in different qualities of grain and other edible vegetables.—Then the price will be determined by the amount of nutritive matter, and both buyer and seller will come to have a common interest in the cultivation of such crops, and the adoption of such methods of husbandry, as will yield the greatest amount of actual aliment, in proportion to the capital, the time, and the labor employed in production.

But the most interesting promise of improvement from a better knowledge of the earth we inhabit, lies in another and more obvious direction. It is remarkable that, while the Roman conquerors of Western Asia, the

Mohammedan invaders of Christendom, and, at a still later date, monks and crusaders, brought from the fertile East, and naturalized in Europe, numerous most valuable products of the vegetable and the animal kingdom, little has been accomplished in recent times in the introduction of plants or animals unknown, to the husbandry of Europe and America. It seems to have been too hastily taken for granted, that these two continents already possessed all the forms of organic life which could be profitably grown or reared in them; and while unbounded labor and expense were incurred in amelioration of familiar products, men had ceased to look elsewhere than at home for the best methods and most valuable objects of agricultural industry. The last half century, which has reduced to comparative insignificance the manufacture of Asia, has, at the same time, better instructed us with regard to the value of the natural productions of those remote and mysterious regions; and we have good ground to believe that our fields are destined to be enriched and enlivened by plants and animals, until now quite strange to us, or but imperfectly made known by descriptive works and scientific collections. France introduced and naturalized the shawl goat of Cashmere and Tibet, more than thirty years since; and several specimens of the yak or mountain ox of central Asia have very lately been brought to that country, in the hope of finding appropriate localities for breeding them in the Pyrenees or other elevated regions of the empire. Madder, which is now cultivated in large quantities with great profit and success in the south of France, was brought from Asia Minor some time in the last century by a Georgian nobleman, who became domiciliated in France. There is an important society, liberally patronized by the French government, which devotes itself to experimenting upon the acclimation of exotics, and its labors are thought to promise very interesting results.

In this country the Thibet goat is said to have succeeded well. The buffalo of the Levant has been brought to South Carolina, where it is supposed he may supply the place of the ox, which does not labor to great advantage in that climate; and our government is now experimenting on a large scale with the dromedary and burden camel. But independently of these and other similar experiments, when we remember that almost every plant which we grow as food for men, except Indian corn and the potato, and all the animals which we rear in the domestic state, besides many tribes of the smaller animated creatures, and of noxious weeds, have been introduced into this continent in the space of three centuries, we cannot but consider it as highly probable that our soil and climate are capable of furnishing localities adapted to a much greater range of vegetable and animal life than we now possess. Some vegetable physiologists have denied the possibility of effecting any such change in the character of plants as to fit them for growth and reproduction in climates liable to greater extremes of heat and cold than those in which the species originated. But our own American experience with maize and the potato, and with numerous plants of tropical and sub-tropical origin, which now grow through a great part of the temperate zone, seems to furnish a satisfactory practical refutation of this doctrine. Indeed, the tomato, which is now thoroughly acclimated, and even spontaneously propagates itself, very often failed to ripen in our Northern states thirty years since, and all the cultivated plants of warmer regions seem to be making some progress to the North.

In attempting the introduction of new objects into our fields and barnyards, we are apt to be discouraged by the difficulty of reconciling foreign organic forms to the new physical conditions which every considerable geographical change implies, and to conclude that because the first crop or the first pair appear to suffer from climatic causes, the species is unsuited to our soil and sky; but though the transplanted plant or animal is seldom so healthy and vigorous as in its native locality, yet in most cases where the contrast is not too violent or too sudden, nature, in the course of a generation or two, accommodates herself to the change of circumstances, and then the progeny very often surpasses the parent stock. The reports of our Patent office are full of valuable suggestions on this important head; and as government has not at present the facilities for extensive experimentation, it is earnestly to be hoped that it may become a subject of special attention from agricultural societies and enlightened and public spirited individuals.

I proposed to add a remark on the spontaneous productions of the European continent. It is a fact, well known to the naturalist, though not obvious to the common observer, that the natural vegetation and animal life of the Old World are seldom or never identical with those of the new, however great the apparent resemblance between them. Misled partly by general similarity of form, and partly by the similarity of names which our forefathers applied to the plants and animals of America, for want of knowing the native appellations, or because they did not notice the specific differences, we are apt to overlook distinctions which to the scientific eye establish a diversity of origin. In corresponding climates, nature produces not identical, but representative species. In the colder regions of Europe, you see the elm, the oak, the birch, the pine and the fir, all bearing so homelike an aspect that you are ready to recognize them as old and familiar acquaintances; but these trees are all, in fact, specifically different from our own. The same law prevails in animated nature, and it may be laid down as a rule subject to few, and those for the most part doubtful, exceptions, that no tree or shrub, or herb, or flower, or grass, or fish, or fowl, or four-footed beast, or creeping thing, is common to both continents, with the exception of such as man, in his wide migrations has transported with him. This points to a radical difference in soil or climate, or both, which doubtless requires a difference in the processes, if not the objects of rural industry. Providence here, as in all our other conditions, makes large demands on the powers of reason and observation implanted in every human breast; and the exercise of these in every relation is peculiarly incumbent upon the American citizen, not merely by reason of his peculiar privileges and the duties thence resulting, but on account of the physical necessities of his position.

The general result, then, of the careful study of European life in all its relations to material things, is that the character of our soil and climate, earth and sky alike, require us to devise for ourselves such adaptations of all industrial pursuits as will bring them best in unison with our peculiar circumstances, and thus to accommodate our rural life to the conditions in which nature has placed us. In the religious, political, civil, and industrial institutions of Europe, God has given us, his last organized great nation, much for attentive study, nothing for blind imitation. All must be more or less modified to harmonize with American nature, and in our general social life, as in each man's private history, we must be emphatically the architects of our own fortunes.

Report, Made under Authority of the Legislature of Vermont, on the Artificial Propagation of Fish

[1857]

In 1856, the Vermont Legislature adopted a resolution requesting that the Governor "enquire into the present state of the discoveries which have been made in relation to the artificial propagation of fish," as well as to consider what the Legislature should do to promote it. This came at a time in Vermont's history when, following a period of rapid forest clearing and construction of dams, many species of fish had declined dramatically or had become extirpated altogether. In addition, it was a time not far removed from an earlier period of fish abundance, so that many residents remembered how it had once been different.

Governor Ryland Fletcher honored this request by asking Marsh, at that time the Vermont State Fish Commissioner, to write a report on the subject. Marsh himself conducted no original research, but rather he summarized the work that had been conducted up to that time, interpreted the summary in light of what it might mean for Vermont and Vermonters, and appended two of the more important papers for the Legislature's consideration. Marsh reflected on artificial fish propagation from both philosophical and environmental perspectives. Philosophically, he felt that increased opportunities for angling by the average Vermonter were, on the whole, a social good in that it would restore to people many positive personal characteristics.

His environmental arguments, on the other hand, were a mix of optimism and pessimism, foresight and limited vision. Among his prescient observations was the fact that, although declines in fish were the result of many causes, such as overharvesting, dams, pollution, and forest clearing, which he felt could not be completely eliminated, aquatic ecosystems could in some measure be restored through the release of fish raised in artificial hatcheries. Marsh also noted that restoration of aquatic systems could not occur without cooperation among all the states in a watershed. Referring to the Connecticut River, whose upper reach forms Vermont's eastern boundary, Marsh noted the necessity of cooperation with the downstream states of Connecticut and Massachusetts if salmon and shad were to be restored. Without such cooperation, there was little Vermont could do.

Despite Marsh's belief that the forces that led to the decline in fish "for the most part . . . cannot be removed or controlled," the past 140 years have seen great strides in promoting the recovery of aquatic ecosystems. Following its nadir in the 1880s, forest cover in Vermont has expanded to its present 76%, and although there are many exemptions granted for agriculture, vegetation along rivers and streams receives special regulatory consideration to protect water quality. Many dams have been retrofitted with fish ladders or removed altogether. Water quality, although still in need of great improvement, has increased in many ways through pollution control and the expansion of forest cover. Fishing regulations for both sportsmen and commercial operations are in place and generally respected. And political units now cooperate to address issues of watershed protection: Connecticut, Massachusetts, New Hampshire, and Vermont cooperate in the management of anadromous fish in the Connecticut River basin, and New York and Vermont though the Lake Champlain Fish and Wildlife Management Cooperative. By 1998, conditions in the Connecticut River had improved to the point that salmon once again, after exactly 200 years, bred in Vermont's tributaries to that river.

In 1857, the Vermont Legislature, following its receipt of Marsh's report, decided not to enact any legislation regarding artificial fish propagation. Fortunately, subsequent governments were of a different mind. The Vermont Fish Commission was established in 1866, the second permanent state conservation agency established in the nation. The state's first fish hatchery was opened at Roxbury in 1891, and although there are still no laws promoting aquaculture, the state's Department of Agriculture has since the early 1990s promoted private aquaculture enterprises. Marsh's report also reached his friend Spencer Baird at the Smithsonian Institution, who used it to promote the early development of federal fish laws.

TO HIS EXCELLENCY, RYLAND FLETCHER, GOVERNOR OF VERMONT:

The Resolution of the General Assembly, in pursuance of which the following Report has been prepared, does not appear to contemplate experiment or original observation upon the natural or artificial breeding of fish, and the report will therefore present such facts only as have been gathered from foreign and American publications on the subject, together with some consideration of a general nature, which may be thought to have a bearing on the proper action of the Legislature in reference thereto.

Man, whether savage or civilized, has a strong passion for the exciting and exhilarating pleasures of the chase, and an irresistible predilection for pursuits which involve the elements of variety, uncertainty, and chance, over the tamer occupations which demand the exercise of regular industry, and offer to their followers not brilliant prizes, but fixed and humble rewards. Many might, therefore, be disposed to question whether the advantages to be derived from the restoration of the quadrupeds, the fowls, and the fish, that once filled the forests, the atmosphere, and the waters, would not be more than counterbalanced by the mischievous influence, which the opportunity of indulging in pleasures so seductive as those of the sportsman would exert upon the habits of our population.

But aside from the obvious impossibility of so multiplying the wild animals of our territory as to affect seriously the habitual pursuits, or the graver interests of our people, it is believed that any possible evil from this source would be more than compensated by collateral advantages, of a character not unlikely in the present state of American society, to be quite overlooked. The people of New England are suffering, both physically and morally, from a too close and absorbing attention to pecuniary interests, and occupations of mere routine. We have notoriously less physical hardihood and endurance than the generation which preceded our own, our habits are those of less bodily activity, the sports of the field, and the athletic games with which the village green formerly rung upon every military and civil holiday, are now abandoned, and we have become not merely a more thoughtful and earnest, but, it is to be feared, a duller, as well as a more effeminate, and less bold and spirited nation. The chase is a healthful and invigorating recreation, and its effects on the character of the sportsman, the hardy physical habits, the quickness of eye, hand, and general movement, the dexterity in the arts of pursuit and destruction, the fertility of expedient, the courage and self-reliance, the half-military spirit, in short, which it infuses, are important elements of prosperity and strength in the bodily and mental constitution of a people; nor is there anything in our political condition, which justifies the hope, that any other qualities than these will long maintain inviolate our rights and our liberties.

The training acquired in the sports of the chase, as exercised in England, has been of great value and importance to those classes of English society which are possessed of the means of participating in it, and in the severe crisis through which the British troops passed in the late Russian war, it proved to be the best preparation for the field and the camp, which it is possible for civil life and an age of peace to afford. In a country like ours, of small landed estates, narrow enclosures, and rugged surface, the chase could never be pursued upon the great scale, which makes it so attractive, and so imposing a sport in England; and it must be admitted that angling and other modes of fishing are under a few circumstances attended with as great moral and physical benefits as the pursuit of the larger quadrupeds, but they are nevertheless analogous in their nature and influences, and as a means of innocent and healthful recreation at least, they deserve to be promoted rather than discouraged by public and even legislative patronage.

But however desirable it might be, in these and other points of view, to repeople the woods and the streams with their original flocks and herds of birds and beasts, and shoals of fish, it is for obvious reasons, impracticable to restore a condition of things incompatible with the necessities and the habits of cultivated social life. The final extinction of the larger wild quadrupeds and birds, as well as the diminution of fish, and other aquatic animals, is everywhere a condition of advanced civilization and the increase and spread of a rural and industrial population. The number of wild animals which have been thus altogether or nearly extirpated in quite recent times is by no means inconsiderable. Within a few centuries, the wolf and the bear, as well as some large animals of the deer family, have utterly disappeared from the British Isles; the wild ox exists only in the parks of one or two great landed proprietors, and the cock of the woods, a magnificent bird of the grouse tribe scarcely smaller than the turkey, formerly abundant in Scotland, had become totally extinct in Great Britain, and has only lately been re-introduced from Sweeden; and the fox has been preserved from extirpation only by a public opinion which exempts him from ordinary agents of destruction, and spares him as the object of a manly sport.

So on the continent of Europe, the beaver is now so rare that he has been forced to relinquish his habits of associated life and action, and has become a solitary animal; the gigantic wild ox of the German and Slavonic states is confined to a single forest in Lithuania, and other large quadrupeds, which abounded in central Europe but four or five centuries since, are now only known by history and tradition.

In like manner the moose, the deer, the catamount, the wolf, the lynx, the beaver, the vast flocks of pigeons and water fowl, and other birds of passage, which bore so important a relation to the nutrition and the sports of our fathers, are now almost unknown to the natural history of Vermont, and zoologists observe that the clearing of the woods and the complete change in the vegetable products of the soil and the insects that feed upon them, have produced corresponding changes in the kinds and numbers of those smaller animals which being neither valuable for their flesh or their peltry, nor obnoxious for their destructive propensities, are regarded with interest by few but the scientific naturalist.

It should be observed, however, that the partial or total disappearance of many of the smaller birds and land animals is not to be ascribed altogether to a diminished supply of their natural food, but in no small degree to the wanton cruelty of youth, which finds pleasure in the torture and death of innocent and defenceless creatures, and to a mistaken prejudice which often ascribes mischievous propensities to particular birds, quadrupeds, and reptiles that in reality, by the destruction of vast numbers of noxious insects, much more than compensate the little injury they inflict upon crops. The insect in all stages, egg, larva, chrysalis, and winged imago, enters largely into the nutriment of birds and the small quadrupeds, and many of these animals which are popularly supposed to be destructive to grass and grain, in fact depend for their sustenance almost wholly upon insect life, and are accordingly useful as protectors, not injurious as destroyers, of the food of man.

But although we must, with respect to our land animals, be content to accept nature in the shorn and crippled condition to which human progress has reduced her, we may still do something to recover at least a share of the abundance which, in a more primitive state, the watery kingdom afforded.

The luxurious and extravagant habits of imperial Rome first introduced the artificial breeding, or at least feeding and fattening of fish, in both salt and fresh water ponds. With the overthrow of that empire, its civilization and its industry, this practice was discontinued, and the art forgotten. But it was revived in the middle ages by the religious observances of the Papal church, which, by determining that fish and certain favorite species of water fowl were not *flesh*, and accordingly not forbidden food at seasons of fasting and mortification, ingeniously contrived to reconcile the indulgence of the palate with the discipline of Lent. To every favorably situated monastic establishment was attached a fish-pond, which not only supplied the tables of the professed during the prescribed fasts, but often yielded a considerable revenue from the sale of fish to worldly penitents. The success of the monks led to the extension of this branch of industry, and large ponds were constructed by laymen, so that in the sixteenth century fish-ponds were an appurtenance of most great estates whether lay or ecclesiastical.

It is well known that in the earlier periods of the history of Vermont, the abundance of fish in the running waters, and more especially in the ponds and lakes of our interior and our borders, was such as to furnish a very important contribution to the nutrition of a population which the cultivated products of the soil were scarcely adequate to sustain. Lake Champlain and the Connecticut, as well as those of their larger tributaries whose course was not obstructed by cascades, abounded in salmon, and after the disappearance of that fish, those important waters, and all the streams and ponds of the interior, long continued to furnish a liberal supply of different species of the trout family, and of other kinds hardly inferior in value. At present, the numbers of the fish in all our waters, as well as of the otter, the mink, the muskrat and the water-fowl that fed on them, are so much reduced, that this branch of the animal kingdom has ceased to possess any pecuniary value in Vermont; and on the contrary the few that remain are popularly regarded as, in an economical point of view, rather a detriment than an advantage, as furnishing a temptation to idleness, not a reward to regular industry. The diminution of the fish is generally ascribed mainly to the improvidence of fishermen in taking them at the spawning season, or in greater numbers at other times than the natural increase can supply. It is believed moreover, and doubtless with good reason, that the erection of sawmills, factories and other industrial establishments on all our considerable streams, has tended to destroy or drive away fish, partly by the obstruction which dams present to their migration, and partly by filling the water with saw dust, vegetable and mineral coloring matter from factories, and other refuse which render it less suitable as a habitation for aquatic life.

It is however probable that other and more obscure causes have had a very important influence in producing the same result. Much must doubtless be ascribed to the general physical changes produced by the clearing and cultivation of the soil. Although we cannot confidently affirm that the total quantity of water flowing over the beds of our streams in a year is greater or less than it was a century since, or that their annual mean temperature has been raised or lowered, yet it is certain that while the spring and autumnal freshets are more violent, the volume of water in the dry season is less in all our water courses than it formerly was, and there is no doubt that the summer temperature of the brooks has been elevated. The clearing of the woods has been attended with the removal of many obstructions to the flow of water over the general surface, as well as in the beds of the streams, and the consequently more rapid drainage of our territory has not been checked in a corresponding degree by the numerous dams which have been erected in every suitable locality. The waters which fall from the clouds in the shape of rain and snow find their way more quickly to the channels of the brooks, and the brooks themselves run with a swifter current in high water. Many brooks and rivulets, which once flowed with a clear, gentle, and equable stream through the year, are now dry or nearly so in the summer, but turbid with mud and swollen to the size of a river after heavy rains or sudden thaws. The general character of our water courses has become in fact more *torrential*, and this revolution has been accompanied with great changes in the configuration of their beds, as well as in the fluctuating rapidity of their streams. In inundations, not only does the mechanical violence of the current destroy or sweep down fish and their eggs, and fill the water with mud and other impurities, but it continually changes the beds and banks of the streams, and thus renders it difficult and often impossible for fish to fulfil that law of their nature which impels them annually to return to their breeding place to deposit their spawn.

The gravelly reach which this year forms an appropriate place of deposit for eggs, and for the nutriment and growth of the fry, may be converted the next season into dry land, or on the other hand, into a deep and slimy eddy. The fish are therefore constantly disturbed and annoyed in the function of reproduction, precisely the function which of all others is most likely to be impeded and thwarted by great changes in the external conditions under which it is performed. Besides this, the changes in the surface of our soil and the character of our waters involve great changes also in the nutriment which nature supplies to the fish, and while the food appropriate for one species may be greatly increased, that suited to another may be as much diminished. Forests and streams flowing through them, are inhabited by different insects, or at least by a greater or less abundance of the same insects, than open grounds and unshaded waters. The young of fish feed in an important measure on the larvae of species which, like the musquito, pass one stage of their existence in the water, another on the land or in the air. The numbers of many such insects have diminished with the extent of the forests, while other tribes, which, like the grasshopper, are suited to the nourishment of full grown fish, have multiplied in proportion to the increase of cleared and cultivated ground. Without citing further examples, which might be indefinitely multiplied, it is enough to say that human *improvements* have produced an almost total change in all the external conditions of piscatorial life, whether as respects reproduction, nutriment, or causes of destruction, and we must of course expect that the number of our fish will be greatly affected by these revolutions.

The unfavorable influences which have been alluded to are, for the most part, of a kind which cannot be removed or controlled. We cannot destroy our dams, or provide artificial water-ways for the migration of fish, which shall fully supply the place of the natural channels; we cannot wholly prevent the discharge of deleterious substances from our industrial establishments into our running waters; we cannot check the violence of our freshets or restore the flow of our brooks in the dry season; and we cannot repeal or modify the laws by which nature regulates the quantity of food she spontaneously supplies to her humbler creatures.

It is therefore not probable that the absolute prevention of taking fish at improper seasons, or with destructive implements, or indeed that any mere protective legislation, however faithfully obeyed, would restore the ancient abundance of our public fisheries, though such measures might no doubt do much to render them somewhat more productive than they at present are, if the legal and moral power of the legislature to enact and enforce appropriate laws on this subject were somewhat greater.

Although the fortieth section of the Constitution of Vermont, which secures to the people of the State certain rights of hunting and fishing, entrusts the General Assembly with a large discretion in the regulation of those rights, yet it is not clear that the Legislature possesses *all* the power required for the complete protection even of an experimental public fish-breeding establishment, and the State certainly at present has title to no suitable localities for such a purpose. Besides this, the habits of our people are so adverse to the restraints of game-laws, which have been found peculiarly obnoxious in all countries that have adopted them, that any *general* legislation of this character would probably be found an inadequate safeguard. But however this may be, the difficulties of a co-operation with other States by concurrent legislation seem, for the present at least, insuperable. The subject is by no means well enough understood to enable us to determine the proper character of a code so comprehensive as to embrace the territory of three or four states, and there is such a difference of local conditions between States, one of which controls the outlet of a great river as well as the entire course of many of its tributaries, and another whose jurisdiction extends but to the water's edge of the upper portion of its current, that the provisions applicable to no one could have little adaptation to the circumstances of the other. The State of Connecticut is in all respects very favorably situated for experimenting upon the restoration of salmon and shad, and whenever that State and Massachusetts shall have adopted protective or promotive systems suited to their respective conditions, it will be the duty and interest of Vermont to resort to such co-operative measures as the interests and circumstances of the State shall seem to require.

It is believed that our main reliance in this, as in all other matters of economical interest, must be upon the enterprise and ingenuity of private citizens, and that until States more advantageously situated for experimentation than Vermont, shall have taken the initiative, our legislative action should be limited to such further protective laws as private establishments may require, and (which is earnestly recommended,) the granting of liberal premiums for judicious and successful private efforts in the restoration and improvement of the fisheries.

In many European countries, where restrictive and prohibitory laws of all sorts are much more rigidly enforced than with us, the preservation of land and aquatic game has been an object of legislation for centuries, but none of these systems have ever been attended with *general* success, and the possessors of great forests and fisheries, whether royal or private, every where depend rather upon guards and enclosures than upon the terror of the law, for the protection of the objects of the chase or the fishery.

Nor does it sufficiently appear that the governmental establishments for fish-breeding in France and elsewhere in Europe have yet accomplished any very important results beyond the supply of spawn to private operators, and, what is of more consequence, the furnishing of satisfactory experimental evidence that the artificial breeding of fish is not only practicable, but may be pursued with advantage as a branch of private industry, requiring less labor, and not more care or skill, than most other rural employments, by any person who possesses a sufficient extent of appropriate territory and water.

There is little which is new in the methods now followed in France, and they are substantially the same as those originally proposed in Germany by Jacobi, and successfully pursued by him and his successors for a century, though it is but lately that they have received the attention their importance merits. That, with such modifications as difference of climate, species, and natural facilities shall require, they will be equally successful with us, there is no ground for doubt, and the effort to introduce them is well worthy of public encouragement.

As has been already remarked, the fattening, and to some extent, the breeding of fish wholly in artificial reservoirs, has been long and widely practiced in Europe, and not unfrequently in this country, but it is not believed that methods, which leave so little to nature can be advantageously pursued on a larger scale. Trout thus grown are so inferior in flavor to fish caught in brooks and mountain lakes, that they can scarcely be recognized as belonging to the same species, but if hatched, protected, and fed during the first year or two in artificial waters, and then dismissed to seek such food as nature provides, they equal in all respects naturally bred fish, and may be greatly multiplied in number, without any diminution in size, or deterioration in quality. The introduction of fish from distant waters, and their naturalization in their new homes is also practicable to an indefinite extent. Thus the gold fish of China, accidentally escaping from artificial reservoirs in this country, breeds and thrives in American rivers; many fish have found their way from the Hudson to the Great Lakes, and from the lakes to the river, since the opening of the New York Canal, and multiplied in both, and it is even said that a gentleman in New York has succeeded in so far changing the natural habits of the shad, that they pass the whole year and freely breed in his fresh water ponds, without returning to the ocean, or having otherwise access to salt water.

The subject of artificial fishbreeding has attracted much attention in other States, and many interesting experiments have been already tried, or are now in progress, in different parts of the Union. Printed accounts of these are readily accessible, and they are therefore not here detailed, but it has been thought expedient to append to this report an abridged translation of an excellent essay by Professor Vogt, of Geneva, in Switzerland, together with extracts from a Report to the Legislature of Massachusetts, and from the Transactions of the Connecticut State Agricultural Society.

It is recommended that a sufficient number of these documents be printed for general distribution in all parts of the State, and it is thought that they, with Fry's complete treatise on Artificial Fish-breeding, published in New York in 1854, and Garlick's Treatise on the artificial propagation of fish, published at Cleveland, Ohio, in 1857, both of which may be easily obtained, together with such experience as a few trials cannot fail to give, will furnish all information necessary to enable any person of ordinary intelligence and possessed of the requisite local facilities (such as clear ponds, or a sufficient extent of the course of a perennial brook), to prosecute this branch of industry with advantage. (It deserves to be noticed, by way of suggesting a caution which it may be important for us to observe, that the forming of large artificial reservoirs, and damming up or otherwise obstructing and diverting the natural flow of water, has in many instances been found injurious to the health of the vicinity by promoting miasmatic exhalations, and that these works have in Europe often seriously impeded the drainage of the soil, and other modes of physical improvement. The tenacity with which the monks adhered to their privileged fisheries, long delayed the execution of that most interesting and remarkable enterprise, the draining and elevation of the bed of the Val di Chiana in Tuscany; and extensive tracts of the richest soil in Sicily are at this moment kept in the condition of barren and pestilential wastes by similar causes.) The amount of care, time and money required for commencing and continuing a moderate breeding establishment in favorable situations, is altogether insignificant, and would not perceptibly increase the labor or the expense of an ordinary farm, while on the other hand, our supply of healthy and agreeable diet might be greatly augmented, and the general prosperity proportionally advanced.

If private persons undertake experiments in the breeding and rearing of fish, whether for scientific investigation or purposes of profit, there is no good reason why industry and capital thus employed should not receive the same protection as the breeding of any other animal, and it is believed that some legislation should be adopted, prescribing the same penalties for the taking of fish in waters which the proprietor has publicly signified his intention of appropriating to his own exclusive use, as for a trespass or a theft committed upon any other personal property.

It is probably too early to attempt the adoption of legislative measures for restoring the primitive abundance of the public waters of Lake Champlain, but when private observation and experiment shall have made the subject more familiar, it is to be hoped that means may be devised for again peopling them with the lake shad (white-fish,) the salmon, the salmon-trout, and numerous other species of fish, which formerly furnished so acceptable a luxury to the rich, and so cheap a nutriment to the poor of Western Vermont, but which now are become almost as nearly extinct as the game that once enlivened our forests.

The Study of Nature

[1860, *Christian Examiner* 68:33–62]

In this article, published in the Unitarian journal *Christian Examiner*, Marsh directly expresses his philosophy that humans do not and should not live in harmony with nature. Nature is an entity that humans must forcefully overcome, as if at war, in order to achieve their greater spiritual condition. Marsh's view emerged from commonly held cultural and religious perspectives of the nineteenth century about the place of humans in the universe and the relative value of European cultural norms.

These views are not, however, widely supported by conservationists today for a variety of reasons. Since Marsh's time, advances in fields such as ecology, anthropology, and comparative religions have indicated that a philosophy of "humans in conflict with nature" is neither sustainable over time nor evidenced by most of the world's cultures. Today we have a greater appreciation for the values of the many cultures that have lived in harmony with their local environments, sometimes for millennia, while still attaining soaring levels of sophistication in art, literature, and philosophy. Further, as the pace of technological development increases, we have come to recognize that the laws of nature are, in fact, excellent models for society. Understanding that energy moves linearly through a system until it is eventually lost as heat and that matter moves in cycles, changing form and location but never being destroyed, has provided the basis for a renaissance in thinking about everything from the design of factories and cities to the design of regional economies. Conservation strategies today seek to re-establish the balance between human activity and ecological processes. For example, the philosophy of ecosystem management, currently the dominant resource management policy of the U.S. federal government, is specifically based on the principle that a balance can be found between ecological needs, social wants, and cultural norms. Furthermore, people who are involved in resource extraction activities, such as farming and forestry, are searching more and more for strategies to minimize their negative ecological consequences in the hope that through balance will come social and ecological sustainability.

Interestingly, environmental historian William Cronon, in his book *Changes in the Land* about colonial New England, demonstrated that even prior to Marsh's time North Americans were deeply conflicted about the relationship between the degree of civilization and the extent to which resources must be wrested from nature. On the one hand, the view that more "primitive" cultures were less civilized because they were spoiled by the bounty of nature, expressed by Marsh in this article, was widely held, and formed part of the ethical justification of land laws that denied to Native Americans ownership of their tribal lands. Unless the land was "worked" and the fruits of nature wrested from it, the land could not be said to be owned. Yet at the same time, the bounty of nature and the ability to find abundant food without effort was a major promotional tool used to encourage people to emigrate to the New World.

The main purpose of Marsh's article, however, was to argue that the study of nature—and by extension, a concern for nature—is not in opposition to Christian religious practice. In this view, Marsh was visionary. As late as 1949, Aldo Leopold, in his book *A Sand County Almanac,* commented that "the proof that conservation has not yet touched [the] foundations of conduct lies in the fact that philosophy and religion have not yet heard of it." Happily, this situation has changed during the last few decades. An increasing number of religious organizations, Christian and non-Christian alike, are collaborating with environmental groups and speaking out on environmental issues. Coalitions of religious organizations—such as the National Religious Partnership for the Environment, the Interfaith Center for Environmental Stewardship, the North American Coalition on Religion and Ecology, and the Center for Religion, Ethics and Social Policy—represent united efforts of congregations of many different denominations that emerge from the same sense of spiritual awe when viewing the natural world as Marsh expressed.

Further, Marsh argues here that the study of nature and landscapes without "the footprints of man" ennobles both art and literature; that there are entire dimensions to these crafts that cannot be approached without a deep appreciation for the beauty and grandeur of nature. One need only consider the poetry of Gary Snyder and Mary Oliver, the fiction of William Faulkner, the paintings of Georgia O'Keeffe, and the photography of Ansel Adams to appreciate the truth of Marsh's argument.

The life of man is a perpetual struggle with external Nature. Her spontaneous and unelaborated products yield him neither sufficient nor appropriate food, nor clothing, nor shelter; and all her influences, if untamed and unresisted, are hostile to his full development and perfect growth, to his physical enjoyments and his higher aspirations, and even to his temporal existence. While obedience to her dictates is the law of all lower tribes of animated being, it is by rebellion against her commands and the final subjugation of her forces alone that man can achieve the nobler ends of his creation. His first conscious voluntary effort is to protect himself against her adverse influences, and the whole premeditated action of the primitive untutored son of Adam is absorbed in appropriating from her scattered stores the raw material which his rude art may convert into the means of sheltering himself from her rigors, and of prolonging an animal existence so low and so barren, that nothing but the instinctive fear of death gives it value. Thus far, the warfare on his part is of a defensive character; but as social life advances, as his relations are multiplied and ramified, the necessities of his position are felt to demand enlarged and varied means of action; he becomes conscious that he is the rightful lord, and Nature the lawful, though unwilling slave; he assumes an aggressive attitude, and thenceforward strives to subdue to his control, and subject to his uses, all her productive and all her motive powers.

His earliest leisure and opportunity for higher pursuits and more refined enjoyments, for the cultivation and exercise of all those powers and faculties that distinguish the human from the brute creation, are won by the subjugation of the inferior animals and the inanimate world,—by compelling unreasoning muscle and unconscious nature to perform those tasks that had otherwise employed and monopolized his entire physical energies; and the extent of his victories over Nature is a measure not only of his civilization, but of his progress in the highest walks of moral and intellectual life.

When a great philosopher characterized man as "Naturæ minister et interpres," he did wrong to his high vocation. True, he can but interpret, not amend, her laws; but when he sinks to be her minister, to make those laws the rule of his life, to mould his action to her bidding, he descends from the sphere of true humanity, abdicates the sceptre and the purple with which the God of nature has invested him, and becomes a grovelling sensualist or a debased idolater. Wherever he fails to make himself her master, he can but be her slave. In this warfare there is no drawn battle. Within, the sensuous or the intellectual must triumph; without, intelligent energy or the brute forces must prevail. In fact, this struggle and this victory constitute so necessary a condition of all other great human achievement, that man rises higher in proportion to the magnitude of the physical inconveniences and wants he successfully combats and finally vanquishes. Humanity has exhibited her loftiest examples of heroism, and wisdom, and virtue, and great exploit, on those soils and in those climates where the earth, with the latent capacity of giving the most, does yet spontaneously yield the least.

Under such circumstances it is that all our powers enjoy their highest stimulus,—the necessity of exerting every energy to secure the means of sustaining even animal life, combined with the certainty that the same efforts, wisely directed and perseveringly continued, will result in crowning our toils with the enjoyment of every good thing that Providence has anywhere placed within the reach of our species.

Universal tradition has planted our ancestral cradle on the highlands of that continent where all the cereal grains, all the textile vegetables, all the important fruits, all the domestic animals, in short, all the regular sources of supply of food and clothing, known to the Old World, appear to have originated; and this under skies so tempered, and on soils so constituted, that the infant life of our race was made up of lessons of industry, patience, self-denial, and self-reliance, which are necessary ingredients of all human power and virtue. The tribes which trace their origin most directly to the primitive stock, and which, in their subsequent migrations, have peopled regions marked by analogous physical conditions, have inherited the largest share of those fundamental excellences, and, if not the earliest in attaining a seeming and specious culture, have extended farthest the realm of human action, of human virtue, and of human thought.

On the other hand, where man is the spoiled child of what are called "more favored climes," where no rude winter stimulates the constructive power to the contrivance and erection of permanent and secure dwellings, where fairest flower and savory fruit and nutritious bulb reward the slightest effort with rich profusion, and an hour of toil insures a week of careless sensual enjoyment,—there man, even of the noblest race, falls almost to an equality with the brute, and remains at that low level, until, as in some tropical regions, increasing numbers, and a consequently inadequate supply for the material wants of life, once more rouse his latent energies, and compel him to assert his right to be regarded as the master, not the parasitic guest, at the table of Nature.

The study of Nature's laws, therefore, a knowledge of her products and her powers, an independence of her influences, a control over her action, is an indispensable means of the first attainment and subsequent extension of high civilization and social improvement, of every form,—in short, of general human progress, as well as of individual culture; and hence this study becomes one of the most obligatory duties, as well as the most imperious necessities, of our condition.

It is a marked and most important distinction between the relations of savage and of civilized man to nature, that, while the one is formed and wrought upon in mental as well as in physical characteristics by the peculiarities of his birthplace and other external natural causes, the other is more or less independent of their action, and his development is accordingly, in a corresponding degree, original and self-determined. Hence in proportion to man's advance in natural knowledge, and his consequent superiority over outward physical forces, is his emancipation from the influence of climatic and other local causes, and the more clearly does he manifest those attributes of his proper humanity which vindicate his claim to be called a being, not a thing,—a special creation living indeed *in* Nature, but not a product of her unconscious action, or wholly subject to her inflexible laws.

When, by human art, the temperate and the torrid zones are made to yield substantially the same products, and supply the same material necessities of life; or when their inhabitants are familiarized with and brought to the use and enjoyment of each other's peculiar advantages, by more rapid and economical means of transport,—when the mountains are pierced with roads and canals, the plains intersected with iron ways,—when island and continent, and opposite shores of broad rivers and broader seas, are virtually joined by swift and sure conveyances;—then local peculiarities and local interests give place to universal and truly human sympathies: men are set free from mere material influences, and, stripped of their national characteristics and their clannish prejudices, the horizon of their philanthropy is widened, they become cosmopolite in sentiment, and feel, emphatically, that the human family has a common interest, and tends to a common destiny.

Wherever such physical improvements as we have here noticed have reached, we find at least the educated and more favored classes alike in manners, in fundamental moral principle, in general intellectual attainment, in their views of the true aims and possible progress of civil society; and the superficial observer, disappointed at meeting the refinements of Europe in an Asiatic saloon, makes this apparent uniformity a cause of complaint against modern civilization, which, as he conceives, has reduced social man everywhere to one dead level of tame and unpicturesque mediocrity. But this is a wholly mistaken view of the actual moral and intellectual condition of civilized humanity. Man, no longer cramped by special and local restraints, no longer moulded by circumscribed and narrow influences, no longer trained to the exclusive imitation of national models, is in fact more freely developed, and, under whatever resemblances of outward form, is more distinctly individualized, than at that not remote period when every nation had its known and recognizable type, in which all individuality was lost. In our time, the local varieties are disappearing, as the general improvement of the race advances; and while men are specifically more alike, they are individually more distinct, more consciously independent and spontaneous in character and action.

These great moral and social changes are no doubt in large proportion to be ascribed to the extended and successful prosecution of the studies we are advocating; but these studies appeal also with irresistible force to those refined sensibilities which constitute the æsthetical part of our nature, and the man of cultivated taste finds them an indispensable auxiliary in the intelligent enjoyment of the visible world around him.

This, indeed, is not the current opinion of hasty inquirers. We are popularly told, that not only has the progress of physical knowledge promoted a materialistic and sensuous philosophy, but that it has tamed the imagination of the bard, vulgarized the phenomena of nature, and dispelled the poetry of common life. We are referred to the Greeks, as a people philosophic, intellectual, and refined to a degree surpassing all that modern civilization has produced, profoundly wise in all immaterial knowledge save that which the Revelation that was denied them alone can teach, supereminently skilful in every branch of creative and of imitative art, and yet acquainted with only the most obvious physical facts, and with none but false and puerile theories where by to explain the *rationale* of them. But this is, to some extent, a mistaken assumption. The difference between the physical knowledge possessed by the Greeks, as well as by some other ancient nations, and that which the moderns have accumulated, consists partly in its form and direction, not altogether in its amount. The laws of statics, for example, though doubtless differently conceived and enunciated, were as well known to the Greeks as to ourselves, and their works show as familiar an acquaintance with nature *in equilibrio* as the most substantial of modern erections. The permanence and stability of their great structures suppose a careful study and a long observation of the laws applicable to bodies at rest, and the curves of the Parthenon imply, not only the employment of a subtile and refined geometry, but a hitherto unsuspected knowledge of the principles of optics. Their most eminent metaphysicians attached high value to every then known branch of the science of number and magnitude, and their philosophy embraced physics in the most extended sense of the term. Learned artistic criticism has shown that the progress of their formative art was dependent upon the advancement of science, and that their highest excellence in sculpture was not attained until they had reached an elevated point in pure mathematics.

But we must here observe what has not been usually sufficiently considered, that the intellectual conceptions, the moral sensibilities, and the poetic imagery of the Greeks, were narrowed and abridged precisely in the direction where their natural knowledge was most imperfect. With dynamics, the laws of force and motion, they had but a very limited acquaintance. They were ignorant of all the greatest facts and the most comprehensive laws of nature; and, though profound in geometry, they can hardly be said to have possessed an arithmetic. A people, however intellectually gifted, but unenlightened by a revealed and spiritualized religion; unacquainted with the vast periods of improved astronomy and cosmogony; without a knowledge of the magnitudes, the distances, the movements, or the velocities of the heavenly bodies, the speed of electricity or light, or even of modern military projectiles; with a numerical system totally unsuited to dealing with great quantities; with a geography that did not extend beyond the temperate zone, and in longitude embraced but a fourth part of the earth's circumference; and with a history whose earliest probable traditions scarcely went back ten centuries,—such a people could elevate itself to none of the grand conceptions which give such dignity and such sublimity to the science of the stars, and even to the special physical history of our own planet; to none of the lofty imagery with which not our poetry alone, but our daily life, is ennobled; to none of those wide views of cosmopolite philanthropy which make this great globe a theatre for the exercise of a universal charity.

To us, accustomed from infancy to an arithmetical notation that enables the eye to seize at a glance the expression of the vastest quantities, and to combine, multiply, and divide those quantities with magical celerity,—familiar with periods extending through myriads of centuries, with distances measured by millions of the earth's diameters, with magnitudes to which our globe is but a vanishing point, with material velocities that ten thousand times surpass the swiftest old poetic notions of the rapidity even of divine motion, with the countless legions of animated existences which the marvellous revelations of the microscope have made known to us, with the terrors of the polar iceberg and the gorgeous luxuriance of tropical vegetation,—to us, thus instructed, it is difficult to imagine an intellect highly cultivated, but yet without the sublime conceptions of extended space, of prolonged duration, of rapid motion, of multiplied numbers, and of earthly grandeur and beauty and power, which modern science has made as it were a part of our mental constitution. But it is certain that these great ideas had no place in the mind of the imaginative Greek. His views of the sensible creation were as humble and as narrow as his theories of the character and attributes of the Divinity. His habitable earth was of less extent than the dominions of a modern Muscovite Czar, his language had no name for any collective number greater than ten thousand, the steeds that wheeled the chariot of the Homeric Jove moved but by leaps that human sight could measure, and his heavens themselves rested on terrestrial pinnacles. The necessarily circumscribed character of his material conceptions could not fail to react on his moral and intellectual nature; and not only the absence of sublime imagery in his poetry, but his want of an enlarged philanthropy, of a humane and enlightened policy in his foreign relations, and of elevated views of the temporal and eternal destiny of man, was, to some extent, the natural consequence of his limited notions of time and space, and number and motion. To the polished Athenian, as to the modern Chinese, every man of a foreign speech was a barbarian and an outlaw, and he conceded even to the alien Greek none of the rights of our common humanity, except as from time to time a temporary and shifting community of interest made it expedient to recognize them.

We can by no means admit, that a familiarity with the truly expounded laws of nature has any tendency to clothe sensuous things in a humbler garb, or to render more prosaic the current of human life. Wherever modern Science has exploded a superstitious fable or a picturesque error, she has replaced it with a grander and even more poetical truth. In the whole range of those mythologies which are built on the apotheosis of mortal heroes, or the deification of the powers of spontaneous nature, in the cosmogonies of the ancient bards, in the warfare of the Gods and the Titans, we find no such theme for ode or anthem as the recent history of scientific research and triumph supplies in abundant profusion. Which is fitter to be celebrated in immortal song, the fiction of a Jupiter launching the forked lightning to avenge a slight offered to a favored mortal, or the true story of the sage philosopher, who, by the aid of a child's toy, forged fetters to chain the thunderbolt? What imagery could be drawn from nature, as known to the ancient world, so grand as that one among the thousand revelations of our astronomy, which tells us that the sun himself, the inexhaustible source of heat and light and life, neither warms nor illuminates nor enlivens the ether-sea in which he floats; but that all created space, save the atmospheric environment of the puny spheres that circle through it, is shrouded in eternal frost, eternal darkness, eternal death? What conception had Aristotle or Plato so lofty as that sublime speculation founded on modern knowledge of the vastness of created space and the progressive motion of light, which tells us that there are, far within the limits of our own feeble vision, points to which the rays reflected from the scenes of man's earliest labors have not yet reached, so that, with sharpened sight and indefinite power of motion, beings with faculties analogous to ours might choose at pleasure stations from which they could follow, with bodily organs, the lost history of our globe, and even now witness the rearing of the pyramids, the founding of the walls of Rome, the battles of Alexander, the triumphs of Cæsar, or the inauguration of Washington! Or how could those philosophers rise to the comprehension of the tremendous truth, pregnant with such vast physical and moral significance, that there exists, not alone in the human conscience or in the omniscience of the Creator, but in external material nature, an ineffaceable, impoverished record, possibly legible even to created intelligences, of every act done, every word uttered, nay, of every wish and purpose and thought conceived by mortal man, from the birth of our first parent to the final extinction of our race; so that the physical traces of our most secret sins shall last until time shall be merged in that eternity of which not science, but religion alone, assumes to take cognizance!

To take a more familiar instance: Which is æsthetically finer, more poetical even, the fable of Æolus and his obedient blasts, or that theory which teaches the origin, the course, the fixed gyrations, and the stable laws of the fickle winds? To the ancient heathen, the tempest and the breeze spoke of nothing but some petty vengeance or some partial favor of a vulgar demon. The modern schoolboy has been taught to find their general origin in the disturbance of the equilibrium of the equatorial atmosphere by the rays of a vertical sun. He sees the heated and rarefied columns of air rushing upward in those fervid regions, charged with the vapors of Amazon and Orinoco, or the light sand-clouds of the Lybian and Arabian deserts, bearing on their wings the minuter insects and the germs of vegetable life, and perfumed with the odors of a tropical flora. He watches them as they now, condensed by the cold of a lofty elevation, pour themselves out over the lower strata of the atmosphere, and, urged forward by new ascending masses, roll across the temperate, and sink down in the frigid zone. Meantime, the vacuum created by their ascent has been filled by the expansion and progression of the adjacent columns, and thus new masses are set in motion by an action extending itself constantly farther and farther through the torpid and heavy atmosphere that surrounds the pole, until the columns first disturbed reach that ultimate goal, begin their retrograde march, and sweep along the surface of the earth with accelerated velocity to their original starting-point, there again to renew the same never-ending circuit. With the theory of this grand and simple movement, our pupil connects the variations of the atmospheric current occasioned by the revolution of the earth, the change of seasons, the tidal action of sun and moon, the alternations of day and night, the inequalities of the earth's surface, and the disturbances produced by the influence of electricity and a thousand other yet obscurer causes; finds order and regularity and law in the action even of the winds, the familiar symbol of capricious change; and comprehends the beneficent results of this ceaseless motion in the distribution of the early and the latter rain, in compensating and equalizing the extremes of temperature, and in dispelling irrespirable gases and noxious miasmata, which else would lower over the face of the globe, and render its surface no longer habitable by man. To the meteorologist, therefore, the dumb and viewless

breeze suggests images, now of tropical luxuriance, and now of polar desolation; speaks in turn of the palm-tree and the iceberg, of equatorial days
shared equally by darkness and by light, and of climes where the entire year
knows but one summer day and one winter night; brings alternate tokens
from zones of sweltering heat, and lands congealed by perpetual frost,—
from kingdoms enlivened by every voice and swarming with every form of
organic life, and from realms of silence, loneliness, and death.

In fact, the observation and knowledge of nature has vastly increased the
wealth of imagery at the command of the poet, and opened to him an inexhaustible mine of the noblest illustration, in veins quite unknown to the ancient world. Verse is fast ridding itself of the flat conventionalities drawn
from old mythologies and erroneous or symbolically expressed physical
theories, and which, though once instinct with life, and even truth as seen
from the point whence they were originally contemplated, have long ceased
to have any significance or force for us, and are, if not unsightly excres-
cences, little better than expletives, in modern poetry. In all references to
natural objects and natural processes, poetry now gives us living images in
place of stereotyped and hollow phrases. The descriptive sketches of
Wordsworth, or Bryant, or Robert Browning, or Tennyson, without parade
of physical science or use of technical nomenclature, make us feel in every
line that here indeed man is the interpreter of nature.

With a system of physics which was scarcely able to refer any one natural effect to its true cause, and which can hardly be said to give the actual
theory of a single material phenomenon, it could not be expected that liter-
ary or pictorial genius should find much matter for truthful illustration in
the works of inanimate nature; and accordingly it is living, self-conscious
being alone that appears with its appropriate local color in the imaginative
or the imitative art of ancient Greece and Rome.

In the history of modern art, we find that no acuteness of unscientific observation could teach the painter truly to represent the most familiar natural
object before him. Scientific analysis has always preceded successful imitation. The study of optics forces itself upon the painter, as essential to the practice of perspective, both linear and aerial, and the relations of different planes,
the more general effects of direct and reflected light and of shade, were
understood so early, that Fra Bartolomeo and Leonardo da Vinci and Becca-
fumi and Dürer, who are all historically known to have carefully investigated
the theory of these laws, attained as great excellence in foreshortening, and all
that belongs to depth and distance, as the ablest of their successors.

The popularization and general diffusion of science, of which Humboldt was the great apostle, has been of infinite service to the cause of art,
because it has brought exact knowledge of at least the external forms and

characteristics of visible objects within the reach of a large class of observers, to whom that knowledge would otherwise have been unattainable; and,
at the same time, it has greatly promoted the advancement of true science
itself, because, by obliging its expounders to clothe it in more attractive
forms, and give it, so to speak, a more picturesque aspect, it has secured the
devotion of a far more numerous host of votaries. We cannot here omit to
perform an act of justice to this greatest of the priesthood of nature, by repelling the guarded insinuations which have ascribed his catholic and generous views of science to the want of a more rigorous mathematical training. The real difficulty in the minds of his timid detractors is not the defect,
but the excess, of his knowledge. A learning which embraced the whole
past history and present phase of every branch of physical research, and
which was moreover graced with the elegances of all literature and dignified with the comprehensive wisdom of all philosophy, cannot but be a
reproach to narrower natures, which see and appreciate truth, not in the
mutual interdependence of apparently unrelated knowledges, but only in
the naked proportions of number and magnitude.

The special labors of Humboldt are most familiar and conspicuous in
the new form which he has given to the study of geography. As taught by
him and his school, it is no longer a fortuitous assemblage of independent
facts and quantities, but its dry details have assumed an organic form and a
human interest, and it has become at once a poetry and a philosophy.

In the genial influence of Humboldt and other kindred spirits, we find
the clearest exemplification of the connection between the progress of science and that of the fine arts, and especially of that new branch of imitation
which owes its present character, if not its specific existence, to an increasing acquaintance and sympathy with nature, the art of landscape-painting.

With rare exceptions, like that of the younger Pliny and some few others, the ancients, in spite of their high æsthetical culture, were, in general,
much more insensible than the moderns to the charms of nature. The love
of landscape and rural beauty has been increased much in proportion to our
familiarity with the different branches of natural knowledge, whose objects
are embraced in every extended view of the earth's surface. Cicero could
travel from Rome to Cilicia, and fill his frequent epistles with selfish trifles,
while scarcely bestowing a glance upon the magnificent scenery, so new and
so strange to Roman eyes, through which his journey led him. For the same
reason, landscape-painting cannot be said to have existed as an independent branch of the pictorial art, and it appears to have been only employed
as a mode of architectural decoration.

A renowned German writer has asserted that Hackert, an artist of the last
century, was the first painter who in his landscapes correctly discriminated

between the different species of trees. The early painters of Germany itself might have taught the critic that this sweeping assertion was an error; but it is nevertheless true, that not only foliage, but all forms of vegetation, were rather conventional than imitative in art, until the diffusion of botanical knowledge had made specific distinctions familiar.

Geological science is of later origin and growth than most other branches of natural knowledge. The common eye is not yet trained to the study and recognition of the inorganic, eruptive, or sedimentary forms and outlines which mark the surface of the earth, and still less to the relations between different heights or formations and the vegetable organisms specially appropriate to them. Hence it is that few landscapes have a true geographical character, and it can hardly be said that the artist has yet appeared who can paint a sand-plain, a shattered lava-current, a natural prairie, or even a rocky pinnacle unclothed with moss or other vegetation. So indispensable is scientific knowledge to the faithful portraiture of nature, that the zoölogist and the botanist find it almost impossible to procure from an unstudied artist, however otherwise skilful, an accurate drawing of the simplest living specimen, or single organ, and the illustrations even of professed works on natural history, though artistic in execution and beautiful to the eye, are not unfrequently wholly defective in the representation of marks of specific difference, or of the general scientific character and distinctive physiognomy of the original.

In like manner, in the rendering of atmospheric phenomena it was only the most palpable features, such as the blue depths, or the shapeless mist, that were seized by the painter, and clouds and skies were conventional and stereotyped, until Howard drew attention to the specific forms of vaporous aggregation, and science began to divine the relations of cause and consequence to which the cumulus, the stratus, and the cirrus, with their various combinations and modifications belong.

The heavenly bodies lie so nearly without the painter's sphere, that they are hardly legitimate objects of pictorial representation. They must, however, be introduced into evening and moonlight landscape, and are often indispensable accessories in allegorical painting. But so inaccurate are the usual notions of artists respecting them, that even so conspicuous an object as the moon is seldom truthfully depicted; and in spite of the familiarity which the observations of six thousand years, and the general diffusion of astronomical knowledge, must be supposed to have given all men with that luminary, one of the most celebrated landscape-painters is said to have represented our satellite in an impossible form or position in no less than three pieces which figured at one exhibition.

The author of "The Modern Painters," the most eloquent if not the soundest of artistical critics, has made landscape-painting the subject of the profoundest study; and if he has not succeeded in establishing all his æsthetical theories, he has, at least, given us the strongest incentive and the best guide to the intelligent non-professional observation of nature which exists in any literature. The most pernicious of Ruskin's errors is the doctrine which, in spite of occasional disclaimers, pervades all his works, that beauty is only for the rich, because costliness and rarity and inherent difficulty are all essential elements of the beautiful in art. The attainment of excellence in any field of human effort is indeed difficult, and always involves persevering, conscientious, and self-sacrificing labor. But this is the price which Providence imposes upon the artist, not a necessary or legitimate ingredient in the pleasure that the spectator derives from beauty and harmony of form and color. With Ruskin, it is a profanation to build a grain-market with clustered columns and sculptured capitals, with groined arches and fretted vaults, to decorate a shop-front with twining arabesques, ornamented friezes, or painted windows, or even to plant flower-beds in the court of a railroad depot or along its track. But is not this the prejudice of a mind to which everything is vulgar that savors of the common cares imposed by providence on all save the members of the favored class to which the critic himself belongs?

Truly considered, the tendency of this familiarization of the eye with beauty in all that environs us is not to degrade art, but to elevate and ennoble common life. The Campanile of Giotto, the Paradise-gates of Ghiberti, the spires of St. Stephen's and of the minsters of Freiburg and Strasbourg, have not become less beautiful because their cost is forgotten, or because they look out forever on busy throngs of men engaged in the trivial concerns of humble humanity; nor was the artistic sense of the Græco-Italians debased, because even their kitchen utensils were modelled and decorated with a taste which our most ornate toys cannot surpass.

Equally mistaken is our theorist in holding that the works of nature are admirable only as the poor life of man has illustrated them, and consequently that the face of creation is an unworthy blank in a vast proportion of the continent we inhabit. Wanting ancient memories, American landscape can have no present beauty, and that which God has created cannot acquire picturesque significance, or rightfully claim to excite human sympathies, till man has consecrated it by his doubtful virtues, his follies, or his crimes!

While Ruskin allows the beauty of the individual tree and rock and flower and grass and moss, and purling stream, and dashing cascade, yet as the Author of nature has grouped them in broad landscape, half revealing, half hiding with lichens, the weather-beaten summit of the mountain,

clothing its flanks with forests, furrowing its slopes with torrents and cata-
racts, burying its foot in the turf of the flowery prairie which stretches beneath it, laving its base with winding rivers, and bounding the view with
the swelling ocean,—this alone has no charms for the intelligent observer,
unless it is stamped with the footprints of man!

This is not a mere misapplication of the principle of association; it is the
error of an eye and a mind habituated to the observation of nature in countries where human action has tamed and modified the primitive outline.
Throughout the Eastern continent, man has everywhere left his visible impress on the face of the material creation. He has stripped the mountain-
sides of their natural vegetable ornament and protection, and laid bare their
surface to the influence of sun and wind and frost and rain, whose action
has denuded the rocks of their earthy covering; thousands of years of assid-
uous toil with plough and harrow and spade and hoe have worn down the
smaller inequalities, reduced the slopes, filled up the slighter depressions,
melted the curves into each other, and thus softened the abruptness of the
natural contour and given the line of vision everywhere a broader and more
flowing sweep. The whole field of view seems as if it had been dressed and
trimmed and smoothed as much with reference to the pleasure of the eye as
to adaptedness to agricultural convenience and use. The intelligent observer here connects the effect with the cause, and this finished regularity of
surface is to him an expressive page of unwritten history, a representative
and an epitome of the industrial life of a hundred generations, and a more
appreciable and impressive proof of the venerable antiquity of Transatlantic
social organization than Egyptian monuments which were already crumbling when Cadmus, the traditional father of Grecian civilization, was cra-
dled. The architectural wonders of antiquity,

> "Those temples, palaces, and tombs stupendous,
> Of which the very ruins are tremendous,"

amaze us with their vastness, their magnificence, the enormous expenditure
of money and of human toil they imply; but they belong to and represent
each its own epoch; they characterize an age, or at most a people. On the
other hand, in the changes which the industry of uncounted successive
generations has produced in the face of European mother-earth, you have
before you, not the accumulated labors of a reign, not of a known historical
dynasty or a race even, but the visible traces of every ploughshare that has
cloven, every spade that has delved the glebe, since the virgin forests
yielded to the axe full thirty centuries ago. Not a sod has been turned, not a
mattock struck into the ground, without leaving its enduring record of the

human toils and aspirations that accompanied the act; and as you mentally follow back the links of this unending chain, you will find them stamped with the impress of multitudes, in comparison with whom the myriads that piled up the Pyramids are but a handful. Every turf is the monument of a hundred lives, and to our eye, accustomed to the verdant and ever-youthful luxuriance of the primitive forest, the very earth of Europe seems decrepit and hoary.

Eyes familiar only with scenery of so artificial a character as this must feel the want of traces of human life in the larger part of our own domain; but, though it must be admitted that the natural configuration of surface is generally less favorable to landscape beauty in the cultivated parts of the United States than in the corresponding portions of the Old World, yet where the elements of that beauty in fact exist, it will be found that they need not the hand of man, or any memorials of his virtues or his vices, to awaken the admiration of every soul that has any true sympathy with creative nature.

It has been insisted that the works of inanimate Nature cannot be a proper object of ideal representation, because it is of the essence of all higher art to ennoble its subject, and that consequently landscape painting, which is purely elective and imitative, not imaginative and creative, is an inferior branch of the profession. There is, however, a more fatal objection to the claims of landscape painting to be ranked among the ideal arts, in the fact that no landscape is a whole, or even a complete part of an organic whole, and therefore it has no type at whose realization the painter can aim; for it is the ability to conceive and divine the typical forms of Nature, to find the point to which her lines converge, but which they never actually attain, that constitutes the essence of creative artistic genius.

Every landscape is merely the fragmentary contingent resultant of unrelated forces successive in time, discordant in action, and tending to no common aim. Certain rocks indeed incline to certain crystalline and aggregate forms, but an assemblage of crystals is not therefore the ideal of a mountain, nor should the painter compose his cliffs of pyramids and prisms. On every existing landscape the destructive forces, too, have exerted their power, and their action is as essential to the production of the general result as that of crystallization itself. The outline of a mountain shot up in jagged peaks by subterranean forces, shattered by earthquake and thunderbolt, and rounded by the slow action of chemical forces, all acting independently of and unconditioned by each other, is its true normal form, and nature tends to produce no other. Each single *organic* kind has its end, and in its production Nature constantly aims to repeat the type which she has chosen because it was best suited to the purposes of its creation. All parts of the man, the brute, the vegetable, are due to the operation of a single law directed to the production of a single whole, to the perfection of which the perfection of all the parts is necessary. The organic being therefore is one, the landscape many. Of all the forces concerned in determining the configuration of terrestrial surface, crystallization and gravitation alone tend to the production of constant form. But the action of these is everywhere controlled by those more powerful agencies to which every landscape owes its most prominent features. A mass of molten rock is thrown up from the bowels of the earth by elastic forces, and poured out to harden upon the surface, or perhaps cooled and crystallized under the pressure of a weight like that which crushed the fabled Titans. The stratum thus formed is injected, disjointed, shattered, elevated, tilted, or overturned by a second convulsion, proceeding from another focus, and acting in a different line of direction. Its summits are lowered, and its angles rounded by wind, and rain, and frost, and atmospheric gases or terrestrial exhalations. A thousand years after, its flanks are rent by an earthquake, and through the fissure pours a deluge from a subterranean reservoir, or from some great lake let loose by the convulsion, which wears down the asperities of its lower regions, fills its depressions with the *débris* of another geological formation, deposits at its base a soil of new ingredients, bears on its current the seeds of a strange vegetation, and confuses its flora with new and inharmonious hues. Again, another catastrophe elevates the sedimentary strata composed of the detritus of its rocks spread out at the bottom of the sea which beats against its cliffs, and hardened by a succession of countless ages. These form extended plains, and, by a second upheaval, a mountain range of different composition and outline. Another convulsion throws up still deeper strata, and produces a third class of mountain, having no community of form or composition with the scenery around it.

In most wide landscapes, all these varied operations of Nature are embraced. No one of the many agencies concerned in them is more a creative or a disturbing force than another, and all alike are constructive causes. In the elements of the landscape, in rock and mountain and plain and sea and river, in climbing vapor and floating cloud and falling shower, there is no ideal type. In all inorganic things, Nature infallibly accomplishes the end she proposes, and needs not the creative imagination of man to portray objects which she has vainly struggled to produce. But throughout all organic life, the process of development is constantly arrested or distorted by disturbing physical causes, harmonious indeed in the operations of creation viewed as a whole, but contingent as respects the individual. Therefore the specific type is never actually realized, and there arises, not merely a pleasing variety, but more or less of actual imperfection. The sculptor or painter, on the other hand, is subject to no disturbances but such as necessarily belong to the inadequacy of his material means, and if he possess that creative gift of which we have spoken, he may conceive and represent forms of a beauty and perfection that Nature has never succeeded in producing.

But the necessity of idealizing organic forms in art arises, not only from the imperfection of Nature's models, but from the demands of our moral constitution, which will not rest without the representation of a purer expression than belongs to mortality. In portrait and historical painting, therefore, it is less the fixed feature and the permanent forms that art strives to improve and elevate, than the fleeting changes of lineament, which give moral expression. Human expression must be ennobled, because man is depraved. His physical nature is distorted, because his moral being is debased.

But no artistic elevation of the works of inanimate Nature is necessary, or even possible. Art cannot ennoble the ocean, when with glassy mirror it gives back the calm heavens, the fleecy clouds, the celestial spheres, or when, convulsed by the tempest, upturned from its uttermost depths and piled up to the sky in rolling masses, it threatens to burst over its eternal landmarks, and shakes the solid earth with its concussion. Nor can the painter idealize or embellish the mist that clings to the side of the mountain, feeding the alpine mosses, or climbs the scarped precipice borne upwards by ascending atmospheric currents, or condenses with the chills of evening and distils drop by drop to form a rill, that soon swells to a torrent, tumbles foaming down the cliffs, glides through the forest, and finally expands into a river freighted with the commerce of nations. The mountain peak piercing the sky with its hoary pinnacles, clothed on its alpine heights with a flora of celestial hues, and subsiding into leafy woods and grassy slopes; the glacier, the iceberg, the cloud pregnant with thunder or gilded by the evening sun,—these admit of no conceivable modifications more grand, more awful, more beautiful, than Nature herself produces, nor is any conventional rule of artistic harmony and composition applicable to bodies which Nature groups by no known law of combination.

But in every single element of landscape, Nature has, if no ideal type, yet her special laws, her general climatic and organic adaptations, which the artist must not violate. Certain slopes and outlines, and consequently certain effects of light and shade, belong to certain geological formations, and particular vegetable forms are assigned to particular soils, climates, and elevations. The painter must not combine alpine scenery with tropical vegetation, he must not bring together upon his canvas things which Nature has put forever asunder.

Landscape painting, then, is but the portraiture of inanimate Nature, and as a moral teacher it can but repeat her lessons. It may reveal to the untravelled eye evidences of the power of the Godhead in the characteristic geologic formations, the sky, or the flora of unvisited regions; in the alp, the glacier, and the volcano; in the overhanging precipices of Sinai, the rock-bound alluvion of the Nile valley, or the burning prairies of the West. It may illustrate that natural law whereby all strong and mighty things are perpetually overcome by weaker forces;—the heavy sea, that floats ponderous navies on its surges, convulsed by the agitation of the light and invisible atmosphere; the solid earth dissolved, and the hardest rocks abraded and transported, by the yielding waters; a royal palace, a triumphal arch, or a monumental pyramid, overthrown by the swelling tendrils of a vine.

But this falls short of the dignity of historical painting, in the same proportion as its subject, unconscious Nature, is beneath the rank of divinely endowed man, and the truest portrayer of spontaneous life and inorganic force can produce no images to rival the artistic exhibition of the grace of motion, the majesty of repose, the expression of fear and remorse, of calm and placid meditation, of joy and hope and resignation, in the perfected forms of human beauty.

Artistic genius has been said to consist in the perfection of the senses. Without admitting that this is a just definition of a prerogative which seems of a nobler order, we may allow that highly cultivated organs are essential to excellence in art, is the best of schools for the eye. The sharpened power of observation thus acquired is almost equivalent to the gift of a new, or rather the recovery of a lost faculty, for the organs of children, like those of savages, possess an acuteness of perception, which, unless assiduously cultivated, disappears, or is greatly weakened, long before the meridian of life. The intelligent rural-bred boy recognizes every bird indigenous to his precinct by the faintest note of its song, or a single flap of its wings in its swiftest flight; he knows the leaf, the flower, the fruit, the seed, the bark, the mode of ramification, the qualities of the wood of every tree and more conspicuous arborescent and herbaceous plant, and distinguishes them with equal certainty, whether clad in full foliage or bared by the frosts of winter. He is familiar with the haunts, the traces, and the habits of every wild quadruped and reptile and fish that frequents the neighboring fields and forests or waters, and in all but name is already a naturalist.

This inborn sympathy with Nature is the source of the highest and most refined enjoyments of which childhood is susceptible, and when duly cherished it tends in a remarkable degree to develop the intelligence and excite the thinking powers of youth. And herein is an exemplification of the remarkable analogies which are often observed between the highest, maturest, most cultivated genius, and the most unsophisticated, most untrained childhood; between the untutored savage and the man of most refined and artificial culture. The necessities, the appetites of infancy and of primitive life, and the pursuit of liberal knowledge, all guide us through paths that lead to the temple of Nature.

To him who has never abjured this native impulse, or who has in maturer life returned to his pristine allegiance to our common mother, the responses of her oracles are the most soothing of external influences, and he is emphatically a wise man who has learned to commune with her in the many tongues in which she speaks to her children. He finds not music only, but profound instruction, in the notes of the song-bird, and the sighing of the pine; for him the voice of the thunder, of the bursting volcano, of the seething ocean, mingle grand and cheering truths with their words of terror. He reads in characters impressed upon the solid rock the historic record of myriads of ages when man was not, and every new field of view unfolds to him evidences of physical revolutions as mighty as the Noachian deluge. If he travels in foreign lands, he sees how, out of the same small stock of primitive materials, Nature has fashioned inorganic forms, sometimes analogous to those of his native soil, sometimes strangely diverse from them; and how, on the other hand, with seeming resemblance of organic life between remote regions, it is yet everywhere so distinct, that, though all living things are representative, each of each, in corresponding climates of the Old and New Worlds, yet neither tree, nor shrub, nor flower, nor grass, nor bird, nor quadruped, nor fish, nor creeping thing, is specifically identical in both. All this diversity of form he recognizes as varied manifestations of the same universal laws. The statute-book of Nature is one, and its provisions embrace every material phenomenon. Its language is universal, and to the initiated it wants no alien interpreter. The student in this faculty has his library and his museum always before him, and needs no cloistered halls, no vast collections, to provide him with teachers or with manuals. He can fill a volume with the history of a spire of grass, and the busy spider, that tapestries with its web the solitary cell of a prison, will furnish a cabinet that may employ the observation of a lifetime. If he cannot visit Arabia or Switzerland, the desert or the glacier, he has still at his command the starry wonders of the mighty firmament, the glories of the rising and the setting sun, the gilded cloud and the rainbow, the mysteries of the falling shower and the drifting snow, the majesty of the thunder and the whirlwind, the secrets of myriad shapes of organic life; and the microscope will reveal to him in the wing-case of a beetle, or the smallest handful of sand, a richer assemblage of gems than ever decked the diadem of an Eastern monarch. Wherever his steps are directed, he finds in the book of Nature the richest of volumes, in her productions the most attractive of social circles, and, to use the words of an eminent divine, he may everywhere enjoy "a happiness surpassing all worldly pleasures, all gifts of fortune,—the happiness of communing with the works of God."

But here the student of physical law must be merged in the Christian philosopher. He must recognize in material things, not the spontaneous products of a self-existing though unconscious organizing power, but the marks and designs of conscious intelligence. He must remember that the pursuit of natural science is not an ultimate but a secondary object, that it is but the stepping-stone to a far higher knowledge, without which it is inept, trifling, valueless; that in its highest forms it reveals but the material relations, the accidents, not the essence of things. From the laboratory, the cabinet, and the botanic garden, he must advance to other halls, and unless he complete the arch of his discipline with the keystone of a spiritual philosophy, he has wasted his materials on a perishable and worthless structure.

The economic value and the lower uses of physical knowledge are too familiar, and the brilliant rewards it holds out to those who most successfully pursue it are too tempting, to make it necessary to appeal to such considerations, and we would rather urge the student to follow it for its own sake, and to seek in it inherent compensations for the toil it may cost him. The great object of accumulating the facts and mastering the principles which together compose the body of natural knowledge is, that we may learn, not how to extract a larger amount of physical good out of the resources of Nature, for this is but an incidental advantage, but how to emancipate ourselves from her power, and make our victories over the external world a vantage-ground to the conquest of the yet more formidable and not less hostile world that lies within.

The applicability of science to the uses of material life is a base and degrading test of its true value; it should be pursued for what it helps us to become, not for that which is enables us to do. Tried by the standard question, "Of what use is it?" in its vulgar acceptation, all that is truly great is well-nigh worthless, and virtue itself becomes vice, when its precepts are followed because they are gainful. The Arctic discoveries of intrepid Kane open no new markets for the cottons of Manchester or Lowell, or the iron of Birmingham or Pittsburgh,—add little to our knowledge of the superficies of our globe, or to our speculative acquaintance with the laws of matter; but who does not feel that the heroic example by which he and his brave companions have refuted the imputation of the effeminacy and degeneracy of the age, and displayed the courage, hardihood, perseverance, discipline, that deified men in the era of the Grecian demigods, is of greater value, in every worthy sense, to American nationality, than if he had brought home from the frozen seas a hundred cargoes of gold and diamonds!

Scholars addicted to the pursuit of religious, ethical, and intellectual philosophy see, or seem to see, dangers to the interests of these grand studies in the zeal with which physical science is now prosecuted, and they believe that a devotion to the cultivation of natural knowledge leads to materialism in philosophy, to scepticism in religion, and to a too sensuous view of the mystery of human existence.

While we condemn an exclusive contemplation of the secret of our being, and of our relations to our Maker and our fellows, from either the spiritual or the physical side, and allow the superior interest and importance of metaphysical and theological science, we cannot admit the fact of any such increased engrossment in material studies as the objection supposes, or any such dangers as are apprehended from their still wider cultivation, provided that each course of inquiry be limited to its own appropriate objects. If, as all Christians and even the wiser heathen admit, the Father of our spirits is also the God of Nature, no one law of his universal code can conflict with any other law, nor can the discovery and disclosure of any truth endanger the reality or the ultimate recognition of any other truth. The firmest conviction that the spiritual life is a mode of being, exempt from the operation of the physical laws of cause and effect, cannot shake the demonstrations by which the dynamic laws of matter are established; and, on the other hand, the conscious certainty of our own spiritual processes and experiences is not to be impugned by arguments founded on external observation, and conclusions deduced from the assumed properties and capabilities of atoms.

The evidences of Christianity are most weakened when the higher proofs are abandoned, and the lower relied upon, and the natural theologies of its friends have done it a greater disservice than the subtlest objections of its assailants. It is a poor Divinity which rests its claims to godhead on the instincts of the beaver or the sagacity of the ant. Nature can teach the existence only of a cause adequate to the effect, and in her works there is nothing which might not be the product of faculties and powers analogous in kind, if in degree sufficiently superior, to our own. The divine attributes with which man, as an immortal, responsible being, are most concerned, and especially the distinctive doctrines of Christianity, are neither suggested, nor can they be established, by proofs drawn from material things. Unlimited power and boundless knowledge do not necessarily imply moral perfection; and though wisdom is usually found allied with goodness, yet they are not one in essence. Spiritual religion must look elsewhere than to the natural world for its evidences; its authority is endangered, not by the revelations of science, but by that weakness of its advocates which appeals to lower and feebler grounds, because, however inconclusive, they are more specious, obvious, and tangible.

We have denied the truth of the assumption, that men are more absorbed in material interests in consequence of the advance of natural knowledge. Doubtless the material action of communities rich in objects of industrial elaboration, or in highly improved mechanical or chemical process, is more conspicuous, from the greatly increased magnitude of the results obtained. The steam-engine, with its half-dozen attendants for operation and supply, does the work of a thousand hands; but does not that very fact imply that some portion at least of those whose labors it performs are thereby relieved from humble toils, and left at liberty for higher callings? Were the rude husbandry, the unscientific routine tillage, the coarse handicrafts, that made every laborer of the Middle Ages an animated machine, less material, in their absolute character or their tendencies, than the refined processes of modern industrial art? Is the boor of agricultural Europe debased, and not rather elevated and humanized, when he is transferred from the clod to the workshop? and does not the guidance of a power-loom imply a more cultivated intelligence than the throwing of a shuttle? The substitution of the wheel for the spindle, and the jenny for the wheel, were steps which benefited society, not merely because they economized human labor and cheapened the comforts of life, but because they tended to develop intelligence, and stimulate thinking, calculating, foreseeing action. The wielding of a tool involves no knowledge of law, no apprehension of a chain of actions intermediate between the cause and the effect; but the simplest machine cannot be operated without a knowledge of its principles, and some intelligence of the laws of Nature on which they rest. The engine-driver, therefore, necessarily thinks, and though occupied with mechanical action, does not himself become assimilated to a machine.

The advancement of natural knowledge and its practical applications have not increased, but on the contrary have sensibly diminished, the proportion of every civilized people actually engaged in mere manual toil. They have withdrawn from purely intellectual pursuits few or none who would otherwise have devoted themselves to those higher studies, but they have exercised the minds and called forth the latent powers of thousands, who, but for their attractions and their rewards, would have remained confined to the humblest walks, and bound to the most material occupations of human life.

A most beneficial effect of the mechanical improvements of modern times is the release of women from a large share of those petty cares, that ceaseless round of household labors, which have hitherto so injuriously affected the health, the temper, and the intellectual life of the female sex. The banishment of the cards and the distaff, the loom and the dye-tub, from the domestic fireside, the multiplied mechanical appliances and the improved culinary arrangements of the present day, have emancipated women from much of that unresting toil, which, commencing with the earliest dawn, and ceasing not, like man's labors, at the going down of the sun, has narrowed and abridged their education, crushed down the cheerful buoyancy and hopefulness of their temperament, prematurely exhausted their physical powers, and made them the drudges, not the helpmates, of the sterner sex.

Hence woman is now free to assume her true position, as not merely a subordinate refining, purifying, and hope-inspiring influence, but as the worthy and equal companion of man, possessing a higher and more generous moral nature than his, with not inferior, though diversely fashioned, intellectual gifts, and therefore as the special elevating element in human society.

It is found that the average intelligence of the whole body politic has been advanced, in the same proportion as the arts founded on a knowledge of the laws and processes of Nature; and while physical science has robbed "divine Philosophy" of none of her votaries, it has given birth, in both sexes, to a numerous and highly trained and disciplined class, which had no existence in stages of society less favorable to the promotion of the pursuits of natural knowledge. The number of studious, thoughtful, truth-seeking men, of the *clerisy*, so to speak, is thereby augmented, and the world as a whole is wiser, and not worse for the introduction of this third estate into the body of our social organization.

But, apart from this particular effect of the elevation of a class of men, or rather the creation of a new order of dignified and intellectual pursuits, it has certainly not been observed, that scholars engaged in scientific research, in detecting the hidden powers and exploring the secret paths of Nature, in binding and loosing her mighty energies, and developing the inexhaustible resources she offers for improving the physical condition of man, are in fact specially disposed to a material philosophy or a sensuous life. The metaphysical German is rapidly advancing to the foremost place in both the investigation of scientific principle and the practical application of physical law; and it is a most significant and important fact, that the present religious reaction in Germany has followed closely upon the increase and diffusion of a taste for natural knowledge, and the advance of mechanical art. Again, if we compare the nations farthest advanced in the knowledge of mechanical principles, and the arts dependent upon them, with those which have remained behind in the march of physical improvement, we shall not find in England, France, or the German States, for example, a smaller proportion of men devoted to the highest walks of intellectual effort and spiritual contemplation, than in untutored Muscovy, or in primitive Italy and Spain. Thus far, then, the apprehensions of the speculative philosopher do not seem warranted by existing facts.

But suppose it were otherwise, and admit that the increased absolute, if not relative, prominence which modern education and prevailing taste give to the study of natural knowledge is an evil; it is still an evil which no voluntary efforts can arrest, and therefore we should strive, not to elude what is inevitable, but to mitigate mischiefs which cannot be altogether averted. The leading pursuits of men, their tastes, their more intellectual recreations even, are directed by external causes over which the individual has no control, and the spring of whose energy lies beyond the reach of society itself. Our abstract opinions with regard to what may be inherently the worthiest objects of pursuit, have but little influence in moulding our tastes, still less in determining the choice of our actual occupations, which depend on the accidental circumstances and interests of birth, fortune, and condition, with a thousand other contingencies that can neither be foreseen nor controlled. It is in vain, therefore, to seek to dissuade men from such lawful course of study and of life as the condition of society from time to time prescribes, and the mass of men, however naturally gifted for abstract metaphysical speculation, are forced to engage in material occupations unfavorable to the repose demanded for the investigation and solution of the highest and most important problems of abstract philosophy. There is perhaps nothing in the course of Providence, nothing in the constitution of man, that authorizes us to believe that the primeval sentence—which, though uttered in the form of a malediction, has hitherto proved, if not directly a blessing, at least a means and condition of all blessings—will be revoked; but if our race is ever generally released from the restraints which physical necessities now impose upon it, and allowed the leisure and the repose which the mastering of the most exalted themes and the climbing of the loftiest summits of human thought require, it will be by such advances in natural science, and such improvements in the dependent arts, as shall enable us everywhere to substitute for the living machine the spontaneous and unwearying forces of inorganic nature. This consummation, so far as it shall be realized, will be due to the successful prosecution of the studies we are defending. The more or less complete emancipation of man from slavery to his own necessities, the ability to devote himself more freely to the highest earthly pursuits, and his consequent indefinite progress in intellectual as well as physical power, is a probable result of the continued advancement of sciences, which have already virtually doubled the span of human life by multiplying our powers and abridging that portion of our days which the supply of our natural wants imperiously claims as its tribute.

Let not then the philosophic philanthropist seek to scare or seduce away the votaries of Nature, or discourage pursuits which claim no rivalry with his own. Let him, on the contrary, by demonstrating the God in and over nature, give a loftier direction to the study of His works, and the investigation of the laws which He has imposed on the material creation. Let him instil into the pupil of science a purer and more reverent spirit, animate him with nobler and more generous aims, and guard him against that sensuous philosophy which makes all spiritual action amenable to material law, and that natural theology which finds in final causes and mutual adaptations, in compensations and expediencies, a system of ethics and divinity so complete as to make a revelation superfluous. With these precepts and these cautions, the disciple may safely be dismissed; and though his investigations may lead him and the world to new views of the relations between natural and spiritual existences, his discoveries in external nature can never endanger truths whose verity was recognized before a physical law had been formulated by man, and will be confessed when our connection with material things shall have ceased forever.

Irrigation: Its Evils, the Remedies, and the Compensations

[February 10, 1874; U.S. Senate, 43rd Congress, 1st session, misc. doc. 55]

This report to the U.S. Commission of Agriculture, written while he was the U.S. Minister to Italy, was Marsh's last published report on an environmental issue. Here he articulated four major views regarding irrigation and water control that were driven by central components of his conservation philosophy. His views brilliantly foreshadowed important issues in water policy in the United States throughout the twentieth century and even to the present day.

First, Marsh felt that since water is such an important resource in some parts of the United States, great care should be taken in developing water delivery systems and policies so that monopolistic private interests cannot gain control of a water supply and destroy the rural middle class. Marsh was particularly sensitive to the connection between access to natural resources and the maintenance of strong rural economies characterized by the dominant presence of small landowners, and he urged the Senate not to place economic efficiency over social justice. For Marsh, the good of the many should be more important than profit for a few. Marsh argued that all waterways should be declared the property of the state, and that no entity should be granted control over waterways in perpetuity. This view was embodied in part in federal water policy, most notably by the passage of the Federal Water Power Act of 1920, which codified federal water authority and established the Federal Power Commission (later restructured as the Federal Energy Regulatory Commission [FERC]). Private licenses for generation of hydropower are granted for only thirty to fifty years, after which time the impact of the hydroelectric power project on the public's interests is assessed and the license reconsidered. The FERC relicensing process has been and will continue to be an important conservation tool for protecting and restoring rivers and streams in the United States.

Second, Marsh argued that irrigation projects would have multiple and far-reaching environmental consequences, including increased incidence of disease, changes in soil chemistry, reduction of stream flow, increased fish mortality, and

flooding due to the development of the reservoirs needed to supply water for irrigation. All these concerns proved to be justified, and most of them remain true today.
For example, irrigation can cause an increase in soil salinity to the point where
crops can no longer be grown. At least one-third of the irrigated lands in the American West currently have some kind of salinity problem, a pattern seen in many
other countries around the world. Similarly, the transport of heavy metals, which
are toxic in high concentrations, by irrigation outflow into downstream wetlands is
a major source of mortality for wetland-dwelling animals.

Third, Marsh believed that an awareness of the problems caused by irrigation
would allow society to develop strategies to avoid the problems wherever irrigation was, in fact, justified. Marsh never interpreted his environmental perspectives—perspectives developed through an appreciation and understanding of science and geography—as arguments for avoiding a technology altogether. Rather,
he clearly felt that recognition of possible environmental problems was an important step in achieving vital social goals, such as food production, in a sustainable
way. Marsh was not against irrigation per se; rather, his focus was on assessing its
overall practicality.

Fourth, Marsh adopted a bioregional perspective in considering the advantages
and pitfalls of irrigation; its benefits and problems would differ from the Southwest to the East, from mountainous areas to plains, and from forested areas to
prairies. He felt that such bioregional sensitivities should be central to federal support for the development—or lack of development—of irrigation, a view that is
receiving increasing support by conservationists today for the assessment of all
technologies.

This report was enormously influential in making the federal government aware
of potential problems associated with the control and distribution of water in the
American arid West and in the development of potential solutions. This report is
credited with beginning the process that led to the establishment of the U.S. Bu-
reau of Reclamation, charged with water control west of the one-hundredth meridian, in 1902.

I. General Considerations.

I. ● Irrigation, or the artificial application of water to growing crops,
is one of the most primitive of agricultural arts, and in the regions
which were the first seats of civilization, reservoirs and canals for collecting,
storing up, and distributing this fertilizing element are among the most
ancient human constructions of which visible remains are still extant.

It is probably chiefly to a sense of the agricultural value of water that we are to ascribe the religious reverence paid to fountains and rivers in the early ages, and which is still traceable as a fading superstition in the names of many holy wells and saints' fountains.

2. The moist atmosphere of the countries whence our population is chiefly derived—the British islands and the Germanic provinces—and the plentiful summer rains of our Atlantic States, have rendered agriculture practicable in all these lands without a resort to the expensive and laborious arrangements involved in rural husbandry, which derives its necessary supply of water not directly from spontaneous rain-fall, but from human art. Hence, except in garden cultivation, and perhaps some other comparatively unimportant branches of agriculture, irrigation has been hitherto practically almost unknown to us.

3. But in many parts of Spanish America artificial watering of the fields has always been as indispensable to successful agriculture as in the Hispanic Peninsula, and our recent acquisition and settlement of a considerable part of the Mexican territory, where the climate makes irrigation a necessity, is now familiarizing us with the practice. Besides this, both in Northern Europe and in the older United States the opinion is fast gaining ground that running, or at least infiltrated water, may often be advantageously supplied to grasses and cultivated vegetables in climates and on soils where precipitation alone was formerly regarded as a sufficient source of moisture for all the field crops. Irrigation is consequently now more or less practiced, and its use is rapidly extending in all the European countries to which I have referred, and there is a strong disposition in the Eastern States of the American Union to test the value of the practice by actual experiment, not merely in market-gardening, but in field-culture, on an extensive scale.

4. There is some danger that, with our characteristic impetuosity and love of novelty, we shall, especially in the comparatively rainless new States and Territories of our vast empire, engage too largely and too inconsiderately in an agricultural process which, in many cases, may be attended with disadvantages more than sufficient to counterbalance the gains from its adoption. At least we may fear that costly arrangements will sometimes be made when simple and less expensive methods would be equally available; and we have reason to apprehend that the acquisition of the control of abundant sources of water by private individuals may often result in the establishment of vested rights and monopolies liable to great abuse, and at the same time calculated to interfere seriously with the adoption of general systems of irrigation.

5. Information in regard to European methods of accumulating and dis-

tributing the water of precipitation, and of flowing springs and rivulets, for agricultural purposes, is readily accessible, and in the practical employment of the system our engineers and the ingenuity of our people will no doubt easily overcome any special difficulties arising from the peculiar geographical and meteorological features of our territory. But the social, legal, sanitary, and financial aspects of the subject in its application to extensive tracts of cultivated land are not familiar to the American public, and for the moment some cautions, of a not altogether obvious nature, are more needed than instructions on points of practical method, or of adaptability to particular branches of agriculture.

6. I propose, then, to point out the evils and difficulties of the practice of irrigation, and to suggest precautions against the occurrence of these evils, and means of palliating them where they are to some extent inevitable.

II. Moral and Social Effects of Irrigation.

7. In a political community where it is now generally admitted that *persons* necessarily have inalienable *rights*, of an extent commensurate with their natural duties and necessities, and that consequently no man or body of men can rightfully use any other man or men simply as a means to a selfish end, it will be allowed that, in the introduction of new systems of industrial or rural occupation, on a scale large enough to affect the rights and interests of whole classes of the population, equal regard should be paid to the good of every class, and few will deny that on all such occasions the moral, social and sanitary consequences of great changes in the habits and employments of large bodies of the people should be considered as of more importance than the merely financial results.

8. In this, as in most other cases of inquiry into questions of political economy, in the present state of that science, we encounter at the outset the great enigma of the right relations between capital and labor—which is really a moral rather than a financial problem—and there are not many instances where those relations are on the whole more unsatisfactory than in the employment of irrigation on the great scale in which it is practiced in many parts of Europe.

9. With an important exception, which I shall notice hereafter, the tendency of irrigation, as a regular agricultural method, is to promote the accumulation of large tracts of land in the hands of single proprietors, and consequently to dispossess the smaller land holders. Where a district, however large, derives its supply of water for irrigation from a single stream or lake,

not of such volume as to be practically inexhaustible, the interests of production require that the husbandry of the entire district be administered on a uniform, or at least on a harmonious, system, and consequently that the control of the source of water supply be vested in a single head. If we suppose a considerable district, with a conveniently accessible water-course of a volume just sufficient to supply it if judiciously distributed, to be owned in severalty by ten different proprietors, it is obvious that each land-holder cannot be allowed to draw off at his pleasure and appropriate to his own use the whole current, or such part of it as may suit his convenience, but the quantity and periods of diversion must be regulated upon some general system established by law, custom or contract for the whole district. The course and capacity of the channels of diversion and of final discharge must be determined by some common principle, and adapted to the branches of husbandry best suited to the soil and climate. It would, in practice, be a matter of extreme difficulty to bring about an agreement between any ten cultivators so situated in regard to the location of such channels, the apportionment of the cost of construction and maintenance, and the assignment of the times of diversion, and the quantities of water to each individual land-holder. Again, these times and quantities must be accommodated to the special crops to be watered, and of course any change in the order or objects of rural husbandry would require a change in the seasons and amount of supply. Hence, the agricultural economy of each farmer must remain substantially fixed and invariable, and even so simple a thing as the rotation of crops would be almost impracticable, because it would be impossible to change the whole system of supply to suit the interests of a single one of the owners. The canals of diversion and distribution once established, the network must consequently remain as immutable as the arteries and veins of the human system, and agricultural progress and improvement would be hopeless.

10. Besides this, the measurement of flowing water, and of course its division between different occupants, are matters of extreme complexity and there would be constant jealousies and dissensions between neighboring claimants in regard to the ascertainment of the quantity rightfully belonging to each, and of the amount actually withdrawn by each from the common source of supply.

11. I have thus far surpassed a case where, as is usual in the United States, streams not navigable are not the property of the state, but are owned either in severalty or by different private proprietors in common. European experience shows, as might be expected from what has just been said, that under such circumstances, as well as where waters belonging to the state are farmed and relet by private individuals, water-rights are a constant source

of gross injustice and endless litigation. The consequence of these inter-minable vexations is that the poorer or more peaceably disposed land-holder is obliged to sell his possessions to a richer or more litigious proprie-tor, and the whole district gradually passes into the hands of a single holder or family or corporation. Hence in the large irrigated plain lands of Europe, real estate is accumulated in vast tracts of single ownership, and farming is conducted on a scale hardly surpassed in England, or even on the boundless meadows and pastures of our own West.

12. There are doubtless considerable economical advantages in the system. The unity of administration tends to increase production as well as to diminish the cost of agricultural operations, but the evils more than counterbalance this advantage. In an often-quoted passage Pliny the elder complained, eighteen hundred years ago, that great farms had been the ruin of Italy, as well as of the tributary territories. He adds in a paragraph not so frequently cited that six land-holders own one-half of the Roman province of Africa, and he thinks it a proof of magnanimity in Pompey that he would never enlarge his farms by buying land adjacent to them.

13. The ruin to which Pliny alludes was not merely from the negligent management of non-resident landlords. He refers rather to the demoraliza-tion of the peasantry in consequence of their abandonment of their native fields and firesides. The small cultivators who sell their paternal acres must either emigrate, and so diminish the resident population, or sink from the class of land-owners to that of hired laborers on the fields which, once their own, are their homes no longer. Having no proprietary interest in the soil they till, no mastership over it, they are, as I have said elsewhere, virtually expatriated, and the middle class, which ought to constitute the true moral as well as physical power of the land, ceases to exist and enjoy a social status as a rural order, and is found only among the trading and industrial popula-tion of the cities.

III. Sanitary Effects of Irrigation.

14. Next in importance to the moral and social aspects of the system we are considering comes the question of the effects of irrigation on the health of the population employing it. In certain branches of agriculture, where water is largely used for irrigation, as in the growing of rice, or in preparing the product for market, as in retting hemp, nothing can be better estab-lished than the fact that the miasmatic exhalations from the soil and the pools are deleterious in the highest degree. The rice-grounds of Lombardy, though principally lying to the north of the forty-fifth degree of latitude,

are almost as destructive to health as those of Georgia and our other Southern States, and the statistics of the increased mortality attending the recent extension of rice culture in northern Italy are truly appalling.

15. But all irrigation, except where the configuration of the surface and the character of the soil are such as to promote the rapid draining of the water, or where special precautions are taken against its influence, is prejudicial to health. In most localities the increased dampness of the atmosphere is injurious to the respiratory system, and in others the exhalations from the watered soil and moistened manures tend powerfully to favor the development of malarious influences, and to aggravate, if not to occasion, febrile diseases.

IV. Physical Evils of Irrigation.

16. From these brief hints at some of the moral, social, and sanitary disadvantages which, under the circumstances supposed, attend the system of husbandry under consideration, I proceed to some notice of the purely physical evils which, in many cases, are inseparable from it.

17. The first and most obvious effect of withdrawing water from its narrow natural channels of flowage, and distributing it over the surface of the earth, is a great increase in the humidity of the soil watered, a like increase in the evaporation from it, and a corresponding reduction of the atmospheric temperature, as in other cases of evaporation. The water imbibed by the earth, which on grounds of slight inclination is generally estimated at about one-seventh of the quantity applied, may not be sufficient to affect the consistence of the soil to a serious degree, but the remaining six-sevenths, so far as not carried off by evaporation, employed to irrigate lands at a lower level, or discharged into running streams or lakes, frequently produce a very prejudicial effect on the soil of adjacent grounds, over which the water flows or into which it percolates. Thus the infiltration of the superfluous water from the rice-grounds of Lombardy, it is said, sometimes renders the lower fields adjacent unfit for any other husbandry to a distance of even miles from the lands flowed for watering the rice.

18. The diversion of brooks and rivers from their natural channels, and the final discharge of the current by remote outlets, tend to deprive the district originally watered by it of their proper supply, and while on the one side considerable tracts of ground are sometimes drenched with superfluous moisture, on the other, water-courses large enough to drive mills and other machinery may be laid dry and their fish destroyed, and even the subterranean conduits from their beds, which fed springs and wells at lower levels, may cease to flow.

19. It has been a general opinion among practical agriculturists that the water employed for irrigation dissolves some of the fertilizing ingredients of the soil to which it is applied, and carries them with it in its flow or percolation over or through the adjacent fields into which it escapes. Hence a higher value has been ascribed to such escape water in its subsequent use than in its original application. This opinion was controverted by Liebig, who taught that none of the material constituents of vegetation were thus abstracted by water, and his views have been confirmed by other observers. Later experiments appear to show that the doctrines of Liebig and his followers are not strictly true, for mineral and vegetable substances which enter more or less into the food of plants have been detected in the water of field-drains and other currents from cultivated soil. Still there is no satisfactory evidence that land is on the whole impoverished by irrigation, though the consistence of the soil may sometimes be affected injuriously by it.

20. The increase of the natural humidity of the soil provokes the growth of aquatic weeds, and although some English writers have asserted that the *marcite* or water-meadows of Lombardy are not infested with these pests, I believe they are kept free from them only by constant weeding. In the rice-fields the extirpation of such plants, which is performed wholly by hand, is perhaps the most laborious and unhealthy of the toil of the cultivator, and in all freely irrigated lands the borders of the channels of distribution are fringed with water-plants, in spite of all efforts to destroy them, and they mark every spot of pasture and meadow surface where the flow or percolation of the water is checked by superficial or underground obstructions.

21. In many localities irrigation cannot be carried on upon a great scale without the construction of large reservoirs for retaining the precipitation of the wet season for use in the dry, and in all Oriental and many European countries such artificial lakes are counted by thousands. Irrigation from reservoirs has most of the general inconveniences which attend other systems, with the additional disadvantages that it is found in practice almost an impossibility so to secure the retaining dams or walls that they do not at length burst their barriers and overwhelm the country below with ruinous desolation. Treatises on hydraulics are full of fearful examples of such calamities, and the construction of works of this sort ought never to be permitted except with guarantees, and under circumstances which promise exceptional security.

22. The quality of the grain, roots and other vegetables cultivated by irrigation is a point of importance, but not hitherto sufficiently investigated. I cannot say that I find the meal of Indian corn or other cereal grains grown in countries where irrigation is generally practiced less sweet or less nutritious than that produced on our unwatered fields. The wheat of Italy is excellent,

and the bread of Andalusia is generally admitted to be both more agreeable to the palate and more nutritious than that of any other country. But water is sparingly applied to maize, and then only in times of drought, and by in-filtration from currents conducted along the furrows, and wheat is hardly watered at all except occasionally in the southernmost provinces of Europe. The grasses of irrigated meadows, especially where, as is apt to be the case when the supply is abundant, water is too freely applied, though luxuriant in growth and of good quality for soiling or feeding green to horned cattle, certainly make hay much less nutritive and less tasteful to stock than that grown on ground watered only by rain. All American travelers find the gar-den vegetables of Continental Europe, peas, beans, tubers, and roots, far less savory than with us, where water is indeed often applied to them, but in much smaller quantities; and it is further observed that though garden-seeds from the United States may produce a single crop of satisfactory quality in a well-watered Italian garden, yet the vegetables grown from the same stock in the following years rapidly deteriorate. Neither I, nor the friends to whom I have given seeds, have been able to obtain anything but stringy pods, even for one season, from American okra-seed, and the lima bean has greatly diminished in productiveness, size, and flavor, and in fact has almost entirely run out, after a few years of partial success in Tuscany.

V. Economical Obstacles to Irrigation.

23. Let us consider the question from a purely economical point of view. Ir-rigation is seldom practicable without a considerable pecuniary outlay in arrangements for collecting and distributing the water, whether derived from springs, lakes, wells, or precipitation. Dams, dikes, artesian borings, common wells, pumping-machinery, reservoirs, aqueducts, siphons, and canals, or some of them, are indispensable whenever irrigation is employed in any cultivation more extensive than ordinary gardening.

24. Besides this, the ground to be watered must be leveled, graded, or scarped, in order to permit either the flow of a current over its whole sur-face, or its gradual absorption and infiltration from the channels into which it is conducted. In new countries, and especially on lands originally wooded, ground is almost always extremely irregular, from the growth of trees and the spread of the roots, which occasion great irregularity of surface, from bowlders, or the cropping out of rock, and from other familiar causes. It is obvious that over the hollows and ridges of such a surface water cannot be evenly and gradually distributed, and measures must be taken to convert the broken curves and ridges of the ground into comparatively uniform

slopes. This, indeed, is effected in a certain degree by the ordinary operations of agriculture, and every farmer knows that an old field is much smoother than a new one. In fact, few things in the long-cultivated regions of the Old World strike the eye of an American farmer more powerfully than the regular slopes and long sweeps of the surface. These outlines are in part the result of special labor devoted to that object; but they are, in a still higher degree, perhaps, the effect of centuries of cultivation under the spade, the hoe, the plow, and the harrow, none of which can be long employed without producing a smoothing down of the original asperities and irregularities of the ground. The natural surface, then, except on alluvial plains, is usually unfavorable to the application of water to grass, or tilled crops grown upon it, and there are many regions in the United States where cultivation has not yet reduced the face of the earth to the necessary regularity, and where, of course, a good deal of labor must be performed in the way of grading before irrigation can be practiced with advantage.

25. All artificial arrangements for irrigation are costly, and, of course, especially in a new country, where much building and other improvement are necessary, so far objectionable. The expenses incurred in them do not belong to the current *annual* account, but the works are generally of a permanent nature, and are therefore so much added to the capital invested. Hence settlers of limited means cannot engage in them, and small land-holding is discouraged. Besides this, the time and attention consumed in watching the canals, and in admitting and shutting off the water are considerable, and where hand labor is so dear as with us, this item would be found a not unimportant addition to the cost of agricultural operations. It is for such reasons that European economists discourage the execution of works for irrigation, unless where an abundant supply of water can be certainly counted on. Boussingault, for instance, states that cheap as is labor in Germany, it is not good economy in that country to construct even the distributing canal and other small subsidiary works, except where at least four inches of water per week can be secured for the whole irrigated surface. This point is important as showing the danger of entering into the system without previous careful inquiry as to the sufficiency of the supply, and this again involves the necessity of experiment sufficiently varied and long continued to determine what, in a given climate and with a given system of agriculture, is a sufficient supply. But I shall return to the point hereafter.

26. There is another suggestion which it is proper to make in estimating the economical value of irrigation, the fact, namely, that in some parts of our own country production is now overabundant, that it needs rather to be repressed than estimated. When the market price of Indian corn is less than the cost of its transportation to the seaboard, and growers can turn it

to no better account than to burn half of it for fuel for distilling the rest, it is evident that the money it costs to raise the surplus yield might be more advantageously spent in creating new facilities for conveying the grain to consumers who require it for higher uses, than in making expensive arrangements for increasing crops which are already so luxuriant as to be a burden rather than a blessing.

27. From all this it will be obvious that considerable evils necessarily attend the practice of field irrigation, and that these would be sensibly felt in its introduction into a country which, taken as a whole, stands in no special need of such a resource for increasing its agricultural production, which has, near its ploughlands and its meadows, no reservoir of Alpine snows to serve as a perpetual source of supply for canals of irrigation, and indeed, though with some important exceptions, no peculiar geographical or climatic adaptation to the system.

VI. Advantages of Irrigation.

28. But my object in pointing out these evils with something of detail has been to inculcate the necessity of caution in attempting a great and general revolution in our agricultural methods, and by no means to discourage careful study of the subject or judicious experiment in appropriate localities. On the contrary, I am well aware that there are extensive territories in our domain where permanently remunerative agriculture is impracticable without irrigation, and indeed I am convinced that there is scarcely any part of our soil where it may not be, at least occasionally, employed with great advantage.

29. The force of the objections I have stated depends much on the physical conditions of the region in which irrigation is employed, and there are whole counties in many of our eastern States where they have little or no application; while in other localities they may, by judicious legislative and economic measures, be almost wholly obviated, and in still others abundantly compensated.

30. Thus in elevated and mountainous districts water is usually abundant, and its sources so numerous that almost any land-holder may secure one or more of them for his own sole use, without clashing with the rights and interests of his neighbor. Hence the division of the soil into comparatively small estates is promoted; for though, in new countries like ours, mountain lands are thinly inhabited and held in large tracts, yet well watered hill pastures gradually rise in value, and these at last become the homes of a comparatively dense population, each of whom is the lord—not of square miles, indeed, but—of acres of productive soil.

31. In such territories irrigation does not injuriously affect the health of the population. Malarious influences are exerted not by flowing or even by freely percolating water. It is only when the fluid stagnates on the surface, or in the soil, that it becomes pernicious. In the hills the inclination promotes the swift flow of water over the ground, or along the canals, and its descent by infiltration is also too rapid to admit it to become a cause of vegetable putrescence. In central and southern Europe almost all the surface of the mountains which has not been swept away by torrents is irrigated through the summer, but fevers and other malarious diseases do not occur in those regions, and they are regarded by many European physicians as especially salubrious even for persons affected with pulmonary complaints.

32. The beneficial effects of irrigation in mountainous countries are not confined solely to the watering of the crops. It checks the too rapid flow of the waters of precipitation, and thus exerts an important geographical if not climatic influence. A large proportion of the water permitted to spread over the surface, and meander along the canals in the upland meadows and pastures, is absorbed by the earth and slowly filtered down, refreshing the roots of the plants it encounters in its passage, until at a somewhat lower level it bursts out in the form of springs. It is a familiar observation in all the older American States that the hills are growing constantly drier and the herbage less abundant, and that the springs which formerly supplied this stock are disappearing. The principal cause of this disastrous change is undoubtedly the destruction of the forests which once clothed the crest of every mountain, and which, it is earnestly to be hoped, will soon be at least partially restored. The replanting of the woods is a slow process, and the continued drying up of the soil is every day rendering it more and more difficult. In the meantime, the introduction of a general system of irrigation at the highest levels where water can still be found, aided by the excavation of simple reservoirs on the hill tops, and at other higher points for retaining the water of rains and melting snows until it can be applied to the surface by canals or absorbed by the earth, would do very much to retard the unfavorable change which is now taking place in the water supply of our mountain farms, and would, at the same time, greatly augment the product of our grass grounds, and often of our plough lands.

33. It has been observed in Europe that draining the soil, either by surface or by under-ground conduits, tends to increase the suddenness and violence of inundations by promoting a too rapid discharge of the waters into river channels. Irrigation in the mountains, or even on the plains, has the contrary effect by retaining much of the water until it can be returned to the atmosphere in the form of vapor. Draining then deranges the harmony of nature by interfering with her methods of maintaining a regular interchange and circulation of

humidity between the atmosphere, the earth, and the sea. Irrigation is in effect a partial return to the economy of our great material parent by regulating that circulation in a manner analogous to her primitive processes.

34. Where springs are numerous, as they usually are in hilly countries, only small and cheap canals, easily accommodated to the accidents of surface, are needed for the diversion of water from its natural channels, to flow over the surface of the ground or to moisten the roots of the grass by infiltration from the artificial water-courses. But the moderate extent and capacity of the necessary canals is not the only advantage of an inclined and undulating surface in the supply of water for the crops. Hilly and winding slopes admit of a simple and efficient mode of irrigation, or rather of a substitute for the practice which is not available on level soils. The method in question has been practiced with success in many parts of the United States, where it is known by the name of *circling*, and it is very highly recommended by all European writers. It consists in horizontally terracing the slopes or even simply furrowing them with the side-hill plow, and leaving the surface permanently in this condition. The rains and melting snow are arrested by the little platforms and ditches thus produced, and gradually sink into the ground instead of running off the surface, and thus supply sufficient moisture for vegetation. It has been found that even in the parched provinces of Southern France soils thus treated produce a vastly increased amount of herbage or of other small crops, and this system, as has been observed in regard to ordinary methods of irrigation, has a collateral advantage of serious importance in countries subject to inundation. The water of precipitation, which soaks into the ground instead of rushing swiftly into the tributaries of great rivers and suddenly swelling them into raging floods, is retained long in the soil and finally carried off by slow subterranean conduction, or restored to the atmosphere through absorption and exhalations by vegetables, or by direct evaporation from the surface, and thus equilibrium is restored. In a large part of our territory, then, and especially in that best suited to the important branch of dairy husbandry, irrigation would not only be unattended with many of the evils which are in some degree inseparable from it on soils of a champaign configuration, but might be introduced at a very moderate cost, and probably with very beneficial results to our agricultural and other social interests.

VII. Duties of Government on This Subject.

35. The expediency of resorting to irrigation as a *general* and regular feature of our entire system of rural husbandry, is a much more complex and difficult

question. Of course I do not now refer to the agriculture of the comparatively rainless zones of our territory, where all cultivation is impracticable without the artificial application of water to growing crops. But even there, urgent as is the necessity of immediate provision for at least a temporary supply of water to lands lately brought, or now about to be brought, under cultivation, there is even greater need of caution and circumspection in the construction of permanent works for the diversion of springs and rivers from their natural channels, and it ought to be the first care of the governments of those territories to see that private individuals or associations do not acquire title by hasty grant or prescriptive rights, by appropriating to their own exclusive use the scanty supply which nature designed for the common benefit of all.

36. The duties of the general and local governments of the United States in regard to the branch of rural economy we are discussing, are by no means confined to the simple protection of natural waters from private encroachment. In the general unfamiliarity of our people with this important subject, we must look to our rulers, both for information on its practical and economical aspects, and for such legislation as shall prevent the greatest amount of evil and secure the greatest amount of good from the introduction of a system so new to us, and which, like all attempts to appropriate to the use of individuals gifts of nature which have long been common to all, must clash with many rooted prejudices, many established customs, and many supposed indefeasible rights.

37. Government ought, then, to take steps for collecting and diffusing the existing knowledge on this subject, and where that knowledge is deficient, to supplement it by encouraging and aiding experiment, and by special inquiry into the physical condition and capabilities, the wants and the means of all our territory where the direct natural supply of water from the heavens is insufficient in quantity, or too irregular in distribution, to satisfy the needs of cultivated vegetation.

38. Much of the practical information needed may be gathered from European experience, and from the study of the methods now employed in those exceptional parts of our territory where irrigation has been long practiced. But the climate and other physical conditions of most of our States, and many of the crops best suited to their soil and sky, are so different from those of the irrigated countries I refer to, that much experiment is needed in order to adapt the fruits of that experience and the application of those methods to soils and crops hitherto cultivated by simpler and more familiar processes. Private experiment in favorable localities may, and doubtless will, do much to throw light on this subject, but the knowledge thus gained will be too local and too special to be of much general value, and we need some

easily accessible preliminary source of instruction to serve as a guide and a safeguard to private enterprise. This can scarcely be furnished, except by either the Federal or State governments, and, in fact, important contributions to our knowledge on this subject have already been published and widely circulated in the reports of the Department of Agriculture at Washington. But what seems to be specially wanted is a series of brief reports of experiments deliberately made by persons of known competence, under Government patronage or sanction, in each geographical section of our country having marked peculiarities of climate, soil, and adaptability to special culture.

39. Such experiments would embrace the investigation of the best methods and seasons of administering water to different crops, and of the quantities required for each; the increase in the quantity of the product from irrigation, as ascertained by actual comparison with similar crops grown without water, but otherwise under similar conditions; the relative quality of watered and unwatered vegetables in point of flavor and of nutritive properties, as well as of richness in saccharine matter, vinous juices, or other important constituents—and this, as I have already observed, has not been sufficiently investigated in Europe; the relative rapidity and period of growth and ripening of watered and unwatered crops; liability to disease or attack by insects; adaptation for drying or other modes of preservation; effect of watering the soil as to danger of injury by frost; accurate computation of the relative cost of cultivation with and without water; and other points which would naturally suggest themselves to experimenters seeking for knowledge rather than for immediate profit.

40. These experiments would not include the construction of great canals of *diversion*, which, as we shall see, ought to be provided for in another way; though the simplest and most economical subsidiary works for the *distribution* of the water should form a subject of investigation.

41. But experiments of this sort would be of comparatively little value unless accompanied, if not preceded, by a hydrographical survey of all the territory which requires or admits of irrigation for agricultural purposes. Even in Italy, where irrigation has been largely practiced for thousands of years, Castellani thought, in the time of the first Napoleon, that such a survey was not only a necessary preliminary to all improvement and all new legislation on the subject, but was indispensable even to the maintenance of the existing system. Such a survey would embrace the quantity of water which can be furnished by the lakes and rivers of each hydrographical basin; the possibility of augmenting the supply by diversion of currents from localities where no agricultural application of them is practicable, or where the supply is in excess of the demand, by artesian borings, common

wells or reservoirs for retaining flood water and melting snow; the ordinary temperature—a point of very great importance; mineral constituents, and quantity and quality of the sedimentary matter of each source; the probable effects of the diversion of currents from their natural channels, as respects the supply for existing or necessary hydraulic works for mechanical purposes, for aqueducts for great cities, for rivulets, springs and wells, at lower levels by infiltration and percolation from river-beds; and the best points for the construction of reservoirs and canals of diversion.

42. Such a survey, or at least something approximate, is an indispensable basis for all sound legislation on the subject; and it will be found, I doubt not, in America, as it has been in most European countries, that the first article of the water-code should be a declaration that all lakes, rivers, and natural water courses are the inalienable property of the State, and that no diversion of water from its natural channels is lawful without the permission of the public authorities. Very probably constitutional amendments would be necessary before such a declaration could be effectual; and independently of this, vested rights, for which compensation ought to be made, may in many cases have been acquired by legislative grant, or by prescription. Perpetual concessions of water-rights to individuals or to corporations, or even grants for more than a very limited number of years, ought to be forbidden by constitutional provision, both on general principles of legislation and because, in consequence of the change of volume in watercourses from the destruction of forests and from other causes, and of the varying numbers and wants of the population, a grant, which at a given period was unobjectionable, may become highly injurious to the public interests ten years later.

43. Not only the natural water-courses, but the principal reservoirs and canals also, ought to belong to the State and be always administered by it. Such works will seldom be properly and securely constructed by private persons or bodies, and the management of them by individuals, and especially by corporations, will always be liable to abuse and gross corruption. No irrigation works, in fact, except for the distribution of water over private grounds, after it has once been withdrawn, under Government supervision, from Government sources of supply, ought to be entrusted to private hands.

44. There are, no doubt, serious objections to the assumption of such burdens and such responsibilities by republican governments, but there are also graver and, as I think, insuperable objections to any other system. Financially, at least, public operations of this sort, if not too precipitately undertaken, might, and probably would be, highly advantageous. The truly stupendous net-work of canals lately constructed in India by the British

Government, taken as a whole, yields a fair rate of interest, and some of the more important branches return annually more than twenty per cent, on their entire cost. The government irrigation works in Italy and France, too have been found highly remunerative as a direct investment. But the financial profits of such works are not by any means to be measured by the income from the rent of the water alone. The vastly-increased production from irrigated lands, by enriching individuals and promoting general prosperity, furnishes a greatly-enlarged base for taxation, and in this, as in most other cases, the best and securest revenue of a State is that which goes to fill the pockets of its citizens. The canal Cavour, lately constructed in Piemonte, and which, it was computed, would water not far from 300,000 acres, was undertaken by a corporation composed, in large proportion, of English stockholders. The managers proved as dexterous financiers as those of the most successful American "rings" and Crédits Mobiliers, and the result was that, though the speculators made fortunes, the shareholders had sunk their capital before the work was completed. The Italian government was obliged to interfere, and finally purchased the canal, and there is no doubt that it will prove very profitable as an immediate investment, and still more so as an indirect source of income, from the increased wealth of the community.

The details of administrative legislation on the subject of water-supply will be found full of difficulty, and experience will doubtless show the necessity of frequent amendment of even the most carefully considered codes which may be adopted. In legislating on a subject so new to us, we cannot expect to give in a day a satisfactory solution to problems which, after centuries of experience, have not yet been fully mastered by the wisdom of the philosophers and lawgivers of Europe. Still, in spite of the difficulty of reconciling American institutions and habits with many of the legal provisions applicable to the material and social conditions of the Old World, I imagine that this branch of European jurisprudence is as well worthy of study and imitation among us as foreign methods of practical irrigation. The literature of European legislation, customary law, and judicial action on this subject is voluminous enough to form a library of itself, and in latter years much has been done to lighten the labor of research on water-questions, and to facilitate the application of the law by legislative codification as well as by the compilation of digests and compends by private jurists. The study of such works, many of which, no doubt, are to be found in the library of the Department of Agriculture, at Washington, would be an almost indispensable preparation for drawing up a code of water-law, and would prevent many errors of legislation both in principle and detail.

45. It is evident that, for constitutional and other obvious reasons, little of the legislation on this subject properly belongs to the Federal Government,

though Congress might take measures to prevent dangerous interference with the natural drainage of what remains of the public lands of the nation, and the acquisition of vested rights in the waters of these lands, as well as generally in the Territories. This might be done either by a general declaration of the right of the nation to the property of such waters, subject to alienation only in favor of State governments, or by special reservation in all land grants in districts where irrigation is, or is likely to become, important. But the regulation and administration of the general water supply, the special legislation respecting use, and against abuse, in the distribution of water, and the prevention of such modes of applying or discharging it as may be injurious to health or to a proper fluid circulation in the territory, must belong to the State governments, which alone can possess the local knowledge necessary to guide the law-making power.

46. In States where irrigation has an actual or a prospective importance as a general agricultural process the Government ought—

To assume the absolute and perpetual ownership of all natural waters except small lakelets and springs or rivulets of a volume barely sufficient for the irrigation of private estates of moderate extent in which they lie or rise;

To provide for a complete hydrographical survey, and, as fast as the necessary information is required and the need of such works is felt, to proceed to the construction of canals of diversion, reservoirs, and feeders, which, like natural rivers and lakes, shall remain forever the property of the State;

To secure the permanence of the springs and streamlets which feed the rivers and the lakes by prohibiting the destruction of the forests around them, and by the plantation of woods in suitable localities, as well as by other available measures:

To prohibit the cultivation of rice or other aquatic crops, and the retting of hemp or other vegetables, except in isolated localities remote from human habitation and routes of travel;

To provide for the drainage or filling up of malarious swamps, and to establish regulations respecting the discharge of water flowing from or over irrigated lands so that it shall not injure lower grounds, or stagnate in or on the soil to the prejudice of the public health;

To prescribe the mode of withdrawal of water from the public sources by canals of distribution for private use, and to determine the method of measurement of the quantity withdrawn;

To make rules for fixing the rates of compensation to the State for the use of the water, and to establish local boards or tribunals with power to

assign the volume and seasons of distribution to the respective landholders within the precincts of the source;

To provide that, in proper cases and on equitable terms the owners or
occupants of lands not bordering on public waters shall have a right of
passage over lands lying contiguous to such waters for constructing and
maintaining canals across them, and of using waste water flowing or percolating from grounds at a higher level.

47. These are the principal points to which the attention of the legislature should be directed; and it is obvious that, with our vast variety of soil,
climate, crops, and geographical surface, no system can be devised which
would be universally applicable. Every State must frame a code suited to
the physical and social conditions of its own territory; and even without
those limits, exceptional provisions adapted to special localities or branches
of agricultural industry will often be requisite.

49. Next to the sanitary regulations, I have no doubt that the leading feature of special State legislation on this subject should be provisions not
merely for securing the rights of small landholders, but for encouraging the
division of the soil into estates or farms of relatively narrow extent. This
may be accomplished, in part at least, by conferring upon such landholders
the privilege of withdrawing a larger proportion of water, or using it at a
lower rate of compensation than that allowed to great proprietors; and further encouragement might advantageously be granted to poorer occupants
who build and inhabit houses upon these lands. The Moors introduced into
their Spanish provinces practical methods of irrigation, and the Spaniards,
a people to whom we do not habitually look for instruction in jurisprudence, have built up on the rude foundation of Moorish water-law a hydraulic code, framed with special reference to this end. A large proportion
of the Spanish peninsula is divided into small parcels and cultivated very
successfully under this system. Many of the provisions of the Moorish law
are still in force in the provinces they occupied, and in some places local tribunals, first instituted by the Moors, having exclusive jurisdiction of water
questions, still exist and hold daily sessions during the season of irrigation,
for summarily trying and deciding, without appeal, all controversies arising
within their precincts respecting water-rights.

VIII. Artificial Methods of Obtaining Water.

50. There are many localities where an adequate supply of water for household purposes and for irrigation may be obtained by simple methods without

resorting to running streams or other public waters. Much land in Italy, Spain, and other countries is irrigated with water drawn from common wells by cheap machinery worked by horse power; artesian and tubular wells are also largely employed for the same purpose; and copious springs may often be reached by driving short tunnels into hill-sides. Another excellent method, practiced with much success in France, is that of conducting the rain and snow-water from hollow slopes of grass-ground of a considerable surface into cisterns, or into filtering receptacles consisting merely of a relatively small extant of sand or porous earth laid over a pavement or bed of clay some four or five feet below the surface. In this way a large proportion of the precipitation received by the slope is retained, and perennial springs are formed at a less expense than is very frequently incurred in conducting water from only a moderate distance. Contributions of this sort deserve encouragement, because they render the farmer independent both of his neighbor and of the public; and even if the first costs of the works is somewhat greater than that of a canal from some source not his own, he will generally be on the whole a gainer by procuring a supply absolutely at his control.

IX. Improving Land and Raising Its Surface by Flooding.

51. I have thus far used the term irrigation in its common and only proper sense, which is the artificial supplying of ordinary water to growing crops. Pure water is a constituent of all vegetable substances, and it also serves as a solvent for other elements of growth, and as a medium for introducing them through the roots into the tissues of plants. It is usually chiefly with reference to these functions of water that means are employed for bringing it within reach of the organs of vegetable nutrition. But water is never found chemically pure, and the current of the limpid spring always contains more or less foreign matter in a gaseous or solid state. This matter generally possesses fertilizing properties, and it consequently both immediately promotes the growth of the plants to which it is applied and enriches the soil by depositing on or in it its extraneous matter not at once taken up by the absorbent action of the plants. This latter advantage is merely incidental, and is never the direct object of what is appropriately called irrigation. But for want of a proper specific word in our English vocabulary, the term irrigation is sometimes applied to a practice, the object of which is not to promote the growth of a single annual or biennial crop, but to enrich permanently the ground to which it is applied, by exposing it to the action of water abounding in the gaseous, the organic, and the mineral elements of

vegetable life, and by depositing on its surface a layer of fertile sedimentary matter, and often, at the same time, to raise considerably the level of the ground. This practice has been extensively and most advantageously employed on the shores of Holland and the adjacent states, as well as in England, and even on our own coast. In these instances the water is salt and is supplied by the tides of the sea. But a similar application of fresh water has long been known, and it has wrought real geographical revolutions in Italy and elsewhere by filling up malarious marshes and other low grounds, and thus at once creating an arable soil and cutting off a dangerous source of fatal disease. In fact, common river meadows, intervales or bottoms, and even vast alluvial plains, as in Egypt, in Assyria, and in Lombardy, which are generally of extraordinary fertility, have been formed by a natural process of this sort, and thus destructive floods, which along the swift tributaries of rivers wash away much valuable soil, make some amends by spreading it out again in broad expanses. It is but lately that anything important has been done in the United States in the way of artificial contrivance for thus utilizing the water of river-floods, and compelling it by dikes and canals to deposit its precious freight at the will and for the convenience of man; but the subject is attracting great attention in Europe, and we in America are beginning to realize the immense benefits which may be derived from a judicious imitation of this, as well as other spontaneous processes of nature. This method of physical improvement is attended with few or none of the evils which are almost inseparable from ordinary irrigation, and it has the important advantage of exercising a beneficial rather than an injurious sanitary action. There is reason to believe that the leveling up of swamps and other low grounds, by retaining on them flood-water long enough to allow it to deposit its sediment, and especially by employing upland streams to wash away waste earth and spread it over morasses and other depressions of surface, would rescue for cultivation and habitation much unproductive and unhealthy soil. Few modes of amelioration of natural conditions are better worthy of public patronage and encouragement, or at least of experiment, than this.

52. The limits of space to which this communication is necessarily restricted have obliged me to content myself with a very superficial view of this important subject, and I cannot enter here at all into financial statistics or other material details. I am, however, thoroughly convinced, after much observation and inquiry, that irrigation may be immensely extended among us with great economical advantage, and that, by reasonable prudence, and above all by a sufficient exercise of moral courage by our rulers, nearly all the evils which ordinarily attend the practice may be avoided, or at least greatly mitigated. We are in a position to protect ourselves and our posterity

by, if I may use so hard a word, *prophylactic* measures—to supply the remedy before the disease manifests itself; whereas in the Old World irrigation had become a widely-spread and deeply-rooted agricultural method before its mischiefs were appreciated, or even suspected, and a preventive policy came too late. European legislation has always been necessarily directed mainly to the combating or compensating of already-existing evils. Ours, by wise precautions, may prevent their occurrence. As I have already said, ample means of information respecting European usage and legislation exist, and the volumes I have specially referred to, though by no means exhaustive, will at least serve as useful introductions to the study of agricultural hydrology.

In a new edition of my *Man and Nature,* now in course of publication at New York, I have discussed with comparative fullness many of the points here barely touched upon; and to that work, and the authorities cited in it, I take the liberty to refer the reader.

Man and Nature; or, Physical Geography as Modified by Human Action

[1864]

Man and Nature was the first major work to argue that humans are a significant force altering the surface of the earth. Prior to its publication, the dominant environmental philosophy was that the earth shaped society. After 1864, the influence of society on shaping the face of the earth was unarguable. Marsh organized the development of his ideas around five main topics—species diversity, forest ecosystems, aquatic ecosystems, desert and dune ecosystems, and large-scale human constructions—each discussed at length in its own chapter. Two themes unite the entire book. The first is that humans are capable of great deeds. Marsh believed strongly in our ability to use technology to solve problems and to make the world a better place from generation to generation. The second theme, however, is that progress, while possible, is not inevitable. Just as humans are capable of great deeds, they are also capable of great destruction. What permits progress is only our careful attention to the laws of nature; we must learn from them and then act on what we learn.

Marsh repeatedly acknowledged that he himself was not a scientist but just an amateur who knew only what he read in the works of others. Yet what he felt was a limitation was, perhaps, an essential part of his great strength. By not restricting himself to one aspect of the question of the relationship between humans and their environment but avidly reading on everything related to this issue and combining this knowledge with his lifetime of observations in the United States, Europe, and the Middle East, Marsh was able to develop an unassailable argument that humans are a powerful geographical force. Although others before Marsh had made the same point, none had done so in a way that transcended their own disciplinary boundaries or that developed so great a vision of the magnitude of human influences on the earth.

Marsh revised *Man and Nature* twice (1874 and 1885), and it was translated into other languages even in his own lifetime. Its impact both within the United States and in other countries was enormous. It fundamentally changed our understanding of humanity's relationship with its own environment. At a practical level, the philosophical perspective he formulated helped in the development of entirely new schools of thought about natural resource management in the United States, especially of forests. Most foresters in the late 1800s and early 1900s credit Marsh for laying the philosophical groundwork for the development of the modern field of forestry in the United States, the creation of forest reserves and a system of national forests in 1891, and for the eventual establishment of the U.S. Forest Service.

Chapter I: Introductory

Marsh began *Man and Nature* with the argument that the collapse of human civilizations in times past was due in large part to a "disregard of the laws of nature." This cautionary theme is one that runs through all of Marsh's writings, as well as much conservation thinking today.

In Marsh's view, human modification of the natural world can be either destructive or productive, depending on the extent to which human actions are guided by improvidence or foresight. This conforms to a utilitarian philosophy that a central value of nature is the provisioning of goods and services to humanity. Marsh believed that nature's ability to provide those goods and services depends on how humans approach natural resource management. This argument is at the heart of current initiatives in sustainable development, which is conceived of as the support of economic development of human cultures through the use of natural resources in ways, based on ecological principles, capable of being sustained indefinitely.

In the introduction to his book, Marsh presented four themes that are still relevant to us today:

• Ecological restoration, which is still considered to be one of the most important tools in conservation.

• Stability of nature, which remains much debated. Marsh had a fairly static view of nature: left undisturbed, landscapes remain unchanged; once disturbed, they succeed over time to a predetermined "climax" condition. Research in many different ecosystem types during the last 50 years tends to support a more dynamic view. Although there may be a general stability to nature (e.g., forest returning to forest), succession following disturbance may result in any of a number of different stable states, each with their own set of dominant species.

• The occurrence of natural disturbance forces in forest ecosystems and the important role they play in altering forest structure. This fact remains central to much conservation work today, as for example the use of prescribed burning in fire-adapted forests and grasslands. As well, Marsh documented examples whereby human modification of landscapes has exacerbated the effects of disturbance and altered the checks and balances of nature.

• The destructive power of humanity. Marsh offered his most forceful argument that "man alone is to be regarded as a destructive power," and contrasted the destructiveness of human actions with those of other species.

Physical Decay of the Territory of the Roman Empire, and of other parts of the Old World

If we compare the present physical condition of the countries of which I am speaking [*countries of the Mediterranean Basin and Middle East*], with the descriptions that ancient historians and geographers have given of their fertility and general capability of ministering to human uses, we shall find that more than one half of their whole extent—including the provinces most celebrated for the profusion and variety of their spontaneous and their cultivated products, and for the wealth and social advancement of their inhabitants—is either deserted by civilized man and surrendered to hopeless desolation, or at least greatly reduced in both productiveness and population. Vast forests have disappeared from mountain spurs and ridges; the vegetable earth accumulated beneath the trees by the decay of leaves and fallen trunks, the soil of the alpine pastures which skirted and indented the woods, and the mould of the upland fields, are washed away; meadows, once fertilized by irrigation, are waste and unproductive, because the cisterns and reservoirs that supplied the ancient canals are broken, or the springs that fed them dried up; rivers famous in history and song have shrunk to humble brooklets; the willows that ornamented and protected the banks of the lesser are gone, and the rivulets have ceased to exist as perennial currents, because the little water that finds its way into their old channels is evaporated by the droughts of summer, or absorbed by the parched earth, before it reaches the lowlands; the beds of the brooks have widened into broad expanses of pebbles and gravel, over which, though in the hot season passed dryshod, in winter sealike torrents thunder; the entrances of navigable streams are obstructed by sandbars, and harbors, once marts of an extensive commerce, are shoaled by the deposits of the rivers at whose mouths they lie; the elevation of the beds of estuaries, and the consequently diminished velocity of the streams which flow into them, have converted thousands of leagues of shallow sea and fertile lowland into unproductive and miasmatic morasses.

Besides the direct testimony of history to the ancient fertility of the regions to which I refer—Northern Africa, the greater Arabian peninsula, Syria, Mesopotamia, Armenia, and many other provinces of Asia Minor, Greece, Sicily, and parts of even Italy and Spain—the multitude and extent of yet remaining architectural ruins, and of decayed works of internal improvement, show that at former epochs a dense population inhabited those now lonely districts. Such a population could have been sustained only by a productiveness of soil of which we at present discover but slender traces; and the abundance derived from that fertility serves to explain how large armies, like those of the ancient Persians, and of the Crusaders and the Tartars in later ages, could, without an organized commissariat, secure adequate supplies in long marches through the territories which, in our times, would scarcely afford forage for a single regiment.

It appears, then, that the fairest and fruitfulest provinces of the Roman Empire, precisely that portion of terrestrial surface, in short, which, about the commencement of the Christian era, was endowed with the greatest superiority of soil, climate, and position, which had been carried to the highest pitch of physical improvement, and which thus combined the natural and artificial conditions best fitting it for the habitation and enjoyment of a dense and highly refined and cultivated population, is now completely exhausted of its fertility, or so diminished in productiveness, as, with the exception of a few favored oases that have escaped the general ruin, to be no longer capable of affording sustenance to civilized man. If to this realm of desolation we add the now wasted and solitary soils of Persia and the remoter East, that once fed their millions with milk and honey, we shall see that a territory larger than all Europe, the abundance of which sustained in bygone centuries a population scarcely inferior to that of the whole Christian world at the present day, has been entirely withdrawn from human use, or, at best, is thinly inhabited by tribes too few in numbers, too poor in superfluous products, and too little advanced in culture and the social arts, to contribute anything to the general moral or material interests of the great commonwealth of man.

Causes of this Decay

The decay of these once flourishing countries is partly due, no doubt, to that class of geological causes, whose action we can neither resist nor guide, and partly also to the direct violence of hostile human force; but it is, in a far greater proportion, either the result of man's ignorant disregard of the laws of nature, or an incidental consequence of war, and of civil and ecclesiastical tyranny and misrule. Next to ignorance of these laws, the primitive source,

the *causa causarum,* of the acts and neglects which have blasted with sterility and physical decrepitude the noblest half of the empire of the Csars, is,
first, the brutal and exhausting despotism which Rome herself exercised
over her conquered kingdoms, and even over her Italian territory; then, the
host of temporal and spiritual tyrannies which she left as her dying curse to
all her wide dominion, and which, in some form of violence or of fraud, still
brood over almost every soil subdued by the Roman legions. Man cannot
struggle at once against crushing oppression and the destructive forces of
inorganic nature. When both are combined against him, he succumbs after
a shorter or a longer struggle, and the fields he has won from the primeval
wood relapse into their original state of wild and luxuriant, but unprofitable
forest growth, or fall into that of a dry and barren wilderness.

Rome imposed on the products of agricultural labor in the rural districts
taxes which the sale of the entire harvest would scarcely discharge; she
drained them of their population by military conscription; she impoverished the peasantry by forced and unpaid labor on public works; she ham-
pered industry and internal commerce by absurd restrictions and unwise
regulations. Hence, large tracts of land were left uncultivated, or altogether
deserted, and exposed to all the destructive forces which act with such energy on the surface of the earth when it is deprived of those protections by
which nature originally guarded it, and for which, in well-ordered husbandry, human ingenuity has contrived more or less efficient substitutes.
(The temporary depopulation of an exhausted soil may be, in some cases, a
physical, though, like fallows in agriculture, a dear-bought advantage.
Under favorable circumstances, the withdrawal of man and his flocks allows the earth to clothe itself again with forests, and in a few generations to
recover its ancient productiveness. In the Middle Ages, worn-out fields
were depopulated, in many parts of the Continent, by civil and ecclesiastical tyrannies, which insisted on the surrender of the half of a loaf already
too small to sustain its producer. Thus abandoned, these lands often relapsed into the forest state, and, some centuries later, were again brought
under cultivation with renovated fertility.) Similar abuses have tended to
perpetuate and extend these evils in later ages, and it is but recently that,
even in the most populous parts of Europe, public attention has been half
awakened to the necessity of restoring the disturbed harmonies of nature,
whose well-balanced influences are so propitious to all her organic offspring, of repaying to our great mother the debt which the prodigality and
the thriftlessness of former generations have imposed upon their successors—thus fulfilling the command of religion and of practical wisdom, to
use this world as not abusing it.

Geographical Influence of Man

. . . [I]t is certain that man has done much to mould the form of the earth's surface, though we cannot always distinguish between the results of his action and the effects of purely geological causes; that the destruction of the forests, the drainage of lakes and marshes, and the operations of rural husbandry and industrial art have tended to produce great changes in the hygrometric, thermometric, electric, and chemical condition of the atmosphere, though we are not yet able to measure the force of the different elements of disturbance, or say how far they have been compensated by each other, or by still obscurer influences; and, finally, that the myriad forms of animal and vegetable life, which covered the earth when man first entered upon the theatre of a nature whose harmonies he was destined to derange, have been, through his action, greatly changed in numerical proportion, sometimes much modified in form and product, and sometimes entirely extirpated.

The physical revolutions thus wrought by man have not all been destructive to human interests. Soils to which no nutritious vegetable was indigenous, countries which once brought forth but the fewest products suited for the sustenance and comfort of man—while the severity of their climates created and stimulated the greatest number and the most imperious urgency of physical wants—surfaces the most rugged and intractable, and least blessed with natural facilities of communication, have been made in modern times to yield and distribute all that supplies the material necessities, all that contributes to the sensuous enjoyments and conveniences of civilized life. The Scythia, the Thule, the Britain, the Germany, and the Gaul which the Roman writers describe in such forbidding terms, have been brought almost to rival the native luxuriance and easily won plenty of Southern Italy; and, while the fountains of oil and wine that refreshed old Greece and Syria and Northern Africa have almost ceased to flow, and the soils of those fair lands are turned to thirsty and inhospitable deserts, the hyperborean regions of Europe have conquered, or rather compensated, the rigors of climate, and attained to a material wealth and variety of product that, with all their natural advantages, the granaries of the ancient world can hardly be said to have enjoyed.

These changes for evil and for good have not been caused by great natural revolutions of the globe, nor are they by any means attributable wholly to the moral and physical action or inaction of the peoples, or, in all cases, even of the races that now inhabit these respective regions. They are products of a complication of conflicting or coincident forces, acting through a long series of generations; here, improvidence, wastefulness, and wanton violence; there, foresight and wisely guided persevering industry. So far as they are purely the calculated and desired results of those simple and familiar operations of agriculture and of social life which are as universal as civilization—the removal of the forests which covered the soil required for the cultivation of edible fruits, the drying of here and there a few acres too moist for profitable husbandry, by draining off the surface waters, the substitution of domesticated and nutritious for wild and unprofitable vegetable growths, the construction of roads and canals and artificial harbors—they belong to the sphere of rural, commercial, and political economy more properly than to geography, and hence are but incidentally embraced within the range of our present inquiries, which concern physical, not financial balances. I propose to examine only the greater, more permanent, and more comprehensive mutations which man has produced, and is producing, in earth, sea, and sky, sometimes, indeed, with conscious purpose, but for the most part, as unforseen though natural consequences of acts performed for narrower and more immediate ends....

Importance and Possibility of Physical Restoration

Many circumstances conspire to invest with great present interest the questions: how far man can permanently modify and ameliorate those physical conditions of terrestrial surface and climate on which his material welfare depends; how far he can compensate, arrest, or retard the deterioration which many of his agricultural and industrial processes tend to produce; and how far he can restore fertility and salubrity to soils which his follies or his crimes have made barren or pestilential. Among these circumstances, the most prominent, perhaps, is the necessity of providing new homes for a European population which is increasing more rapidly than its means of subsistence, new physical comforts for classes of the people that have now become too much enlightened and have imbibed too much culture to submit to a longer deprivation of a share in the material enjoyments which the privileged ranks have hitherto monopolized.

To supply new hives for the emigrant swarms, there are, first, the vast unoccupied prairies and forests of America, of Australia, and of many other great oceanic islands, the sparsely inhabited and still unexhausted soils of Southern and even Central Africa, and, finally, the impoverished and half-depopulated shores of the Mediterranean, and the interior of Asia Minor and the farther East. To furnish to those who shall remain after emigration shall have conveniently reduced the too dense population of many European states, those means of sensuous and of intellectual well-being which are styled "artificial wants" when demanded by the humble and the poor, but are admitted to be "necessaries" when claimed by the noble and the rich, the soil must be stimulated to its highest powers of production, and man's utmost ingenuity and energy must be tasked to renovate a nature drained, by his improvidence, of fountains which a wise economy would have made plenteous and perennial sources of beauty, health, and wealth.

In those yet virgin lands which the progress of modern discovery in both hemisphere has brought and is still bringing to the knowledge and control of civilized man, not much improvement of great physical conditions is to be looked for. The proportion of forest is indeed to be considerably reduced, superfluous waters to be drawn off, and routes of internal communication to be constructed; but the primitive geographical and climatic features of these countries ought to be, as far as possible, retained.

Stability of Nature

Nature, left undisturbed, so fashions her territory as to give it almost unchanging permanence of form, outline, and proportion, except when shattered by geologic convulsions; and in these comparatively rare cases of derangement, she sets herself at once to repair the superficial damage, and to restore, as nearly as practicable, the former aspect of her dominion. In new countries, the natural inclination of the ground, the self-formed slopes and levels, are generally such as best secure the stability of the soil. They have been graded and lowered or elevated by frost and chemical forces and gravitation and the flow of water and vegetable deposit and the action of the winds, until, by a general compensation of conflicting forces, a condition of equilibrium has been reached which, without the action of man, would remain, with little fluctuation, for countless ages.

We need not go far back to reach a period when, in all that portion of the North American continent which has been occupied by British colonization, the geographical elements very nearly balanced and compensated each other. At the commencement of the seventeenth century, the soil, with insignificant exceptions, was covered with forests; and whenever the Indian, in consequence of war or the exhaustion of the beasts of the chase, abandoned the narrow fields he had planted and the woods he had burned over, they speedily returned, by a succession of herbaceous, arborescent, and arboreal growths, to their original state. (I do not here speak of the vast prairie region of the Mississippi valley, which cannot properly be said ever to have been a field of British colonization; but of the original colonies, and their dependencies in the territory of the present United States, and in Canada. It is, however, equally true of the Western prairies as of the Eastern forest land, that they had arrived at a state of equilibrium, though under very different conditions.) Even a single generation sufficed to restore them almost to their primitive luxuriance of forest vegetation. (The great fire of Miramichi in 1825, probably the most extensive and terrific conflagration recorded in authentic history, spread its ravages over nearly six thousand square miles, chiefly of woodland, and was of such intensity that it seemed to consume the very soil itself. But so great are the recuperative powers of nature, that, in twenty-five years, the ground was thickly covered again with trees of fair dimensions, except where cultivation and pasturage kept down the forest growth.) The unbroken forests had attained to their maximum density and strength of growth, and, as the older trees decayed and fell, they were succeeded by new shoots or seedlings, so that from century to century no perceptible change seems to have occurred in the wood, except the slow, spontaneous succession of crops. This succession involved no interruption of growth, and but little break in the "boundless contiguity of shade;" for, in the husbandry of nature, there are no fallows. Trees fall singly, not by square roods, and the tall pine is hardly prostrate, before the light and heat, admitted to the ground by the removal of the dense crown of foliage which had shut them out, stimulate the germination of the seeds of broad-leaved trees that had lain, waiting this kindly influence, perhaps for centuries. Two natural causes, destructive in character, were, indeed, in operation in the primitive American forests, though, in the Northern colonies, at least, there were sufficient compensations; for we do not discover that any considerable permanent change was produced by them. I refer to the action of beavers and of fallen trees in producing bogs, and of smaller animals, insects, and birds, in destroying the woods. Bogs are less numerous and extensive in Northern States of the American union, because the natural inclination of the surface favors drainage; but they are more frequent, and cover more ground, in the Southern States, for the opposite reason. They generally originate in the checking of watercourse by the falling of timber, or of earth and rocks, across their channels. If the impediment thus created is sufficient to retain a permanent accumulation of water behind it, the trees whose roots are overflowed soon perish, and then by their fall increase the obstruction, and, of course, occasion a still wider spread of the stagnating stream. This process goes on until the water finds a new outlet, at a higher level, not liable to similar interruption. The fallen trees not completely covered by water are soon overgrown with mosses; aquatic and semi-aquatic plants propagate themselves, and spread until they more or less completely fill up the space occupied by the water, and the surface is gradually converted from a pond to a quaking morass. The morass is slowly solidified by vegetable production and deposit, then very often restored to the forest condition by the growth of black ashes, cedars, or, in southern latitudes, cypresses, and other trees suited to such a soil, and thus the interrupted harmony of nature is at last reëstablished.

I am disposed to think that more bogs in the Northern States owe their origin to beavers than to accidental obstructions of rivulets by wind-fallen or naturally decayed trees; for there are few swamps in those States, at the outlets of which we may not, by careful search, find the remains of a beaver dam. The beaver sometimes inhabits natural lakelets, but he prefers to owe his pond to his own ingenuity and toil. The reservoir once constructed, its inhabitants rapidly multiply, and as its harvests of pond lilies, and other aquatic plants on which this quadruped feeds in winter, become too small for the growing population, the beaver metropolis sends out expeditions of discovery and colonization. The pond gradually fills up, by the operation of the same causes as when it owes its existence to an accidental obstruction, and when, at last, the original settlement is converted into a bog by the usual processes of vegetable life, the remaining inhabitants abandon it and build on some virgin brooklet a new city of the waters.

In countries somewhat further advanced in civilization than those occupied by the North American Indians, as in mediæval Ireland, the formation of bogs may be commenced by the neglect of man to remove, from the natural channels of superficial drainage, the tops and branches of trees felled for the various purposes to which wood is applicable in his rude industry; and, when the flow of the water is thus checked, nature goes on with the processes I have already described. In such half-civilized regions, too, windfalls are more frequent than in those where the forest is unbroken, because, when openings have been made in it, for agricultural or other purposes, the entrance thus afforded to the wind occasions the sudden overthrow of hundreds of trees which might otherwise have stood for generations, and thus have fallen to the ground, only one by one, as natural decay brought them down. Besides this, the flocks bred by man in the pastoral state, keep down the incipient growth of trees on the half-dried bogs, and prevent them from recovering their primitive conditions.

Young trees in the native forest are sometimes girdled and killed by the smaller rodent quadrupeds, and their growth is checked by birds which feed on the terminal bud; but these animals, as we shall see, are generally found on the skirts of the wood only, not in its deeper recesses, and hence the mischief they do is not extensive. The insects which damage primitive forests by feeding upon products of trees essential to their growth, are not numerous, nor is their appearance, in destructive numbers, frequent; and those which perforate the stems and branches, to deposit and hatch their eggs, more commonly select dead trees for that purpose, though, unhappily, there are important exceptions to this latter remark. (The locust insect, *Clitus pictus,* which deposits its eggs in the American locust, *Robinia pseudacacia,* is one of these, and its ravages have been and still are most destructive to that very valuable tree, so remarkable for combining rapidity of growth with strength and durability of wood. This insect, I believe, has not yet appeared in Europe, where, since the so general employment of the *Robinia* to clothe and protect embankments and the scarps of deep cuts on railroads, it would do incalculable mischief. As a traveller, however, I should find some compensation for this evil in the destruction of these acacia hedges, which as completely obstruct the view on hundreds of miles of French and Italian railways, as the garden walls of the same countries do on the ordinary roads.) I do not know that we have any evidence of the destruction or serious injury of American forests by insects, before or even soon after the period of colonization; but since the white man has laid bare a vast proportion of the earth's surface, and thereby produced changes favorable, perhaps, to the multiplication of these pests, they have greatly increased in numbers, and, apparently, in voracity also. Not many years ago, the pines on thousands of acres of land in North Carolina, were destroyed by insects not known to have ever done serious injury to that tree before. In such cases as this and others of the like sort, there is good reason to believe that man is the indirect cause of an evil for which he pays so heavy a penalty. Insects increase whenever the birds which feed upon them disappear. Hence, in the wanton destruction of the robin and other insectivorous birds, the *bipes implumis,* the featherless biped, man, is not only exchanging the vocal orchestra which greets the rising sun for the drowsy beetle's evening drone, and depriving his groves and his fields of their fairest ornament, but he is waging a treacherous warfare on his natural allies.

(In the artificial woods of Europe, insects are far more numerous and destructive to trees than in the primitive forests of America, and the same remark may be made of the smaller rodents, such as moles, mice, and squirrels. In the dense native wood, the ground and the air are too humid, the depth of shade too great for many tribes of these creatures, while near the natural meadows and other open grounds, where circumstances are otherwise more favorable for their existence and multiplication, their numbers are kept down by birds, serpents, foxes, and smaller predacious quadrupeds. In civilized countries, these natural enemies of the worm, the beetle and the mole, are persecuted, sometimes almost exterminated, by man, who also removes from his plantations the decayed or wind-fallen trees, the shrubs and underwood, which, in a state of nature, furnished food and shelter to the borer and the rodent, and often also to the animals that preyed upon them. Hence the insect and the gnawing quadruped are allowed to increase, from the expulsion of the police which, in the natural wood, prevent their excessive multiplication, and they become destructive to the forest because they are driven to the living tree for nutriment and cover. The forest of Fontainebleau is almost wholly without birds, and their absence is ascribed by some writers to the want of water, which, in the thirsty sands of that wood, does not gather into running brooks; but the want of undergrowth is perhaps an equally good reason for their scarcity. In a wood of spontaneous growth, ordered and governed by nature, the squirrel does not attack trees, or at least the injury he may do is too trifling to be perceptible, but he is a formidable enemy to the plantation.)

In fine, in countries untrodden by man, the proportions and relative positions of land and water, the atmospheric precipitation and evaporation, the thermometric mean, and the distribution of vegetable and animal life, are subject to change only from geological influences so slow in their operation that the geographical conditions may be regarded as constant and immutable. These arrangements of nature it is, in most cases, highly desirable substantially to maintain, when such regions become the seat of organized commonwealths. It is, therefore, a matter of the first importance, that, in commencing the process of fitting them for permanent civilized occupation, the transforming operations should be so conducted as not unnecessarily to derange and destroy what, in too many cases, it is beyond the power of man to rectify or restore.

Restoration of Disturbed Harmonies

In reclaiming and reoccupying lands laid waste by human improvidence or malice, and abandoned by man, or occupied only by a nomade or thinly scattered population, the task of the pioneer settler is of a very different character. He is to become a co-worker with nature in the reconstruction of the damaged fabric which the negligence or the wantonness of former lodgers has rendered untenantable. He must aid her in reclothing the mountain slopes with forests and vegetable mould, thereby restoring the fountains which she provided to water them; in checking the devastating fury of torrents, and bringing back the surface drainage to its primitive narrow channels; and in drying deadly morasses by opening the natural sluices which have been choked up, and cutting new canals for drawing off their stagnant waters. He must thus, on the one hand, create new reservoirs, and, on the other, remove mischievous accumulations of moisture, thereby

equalizing and regulating the sources of atmospheric humidity and of flowing water, both which are so essential to all vegetable growth, and, of
course, to human and lower animal life.

Destructiveness of Man

Man has too long forgotten that the earth was given to him for usufruct
alone, not for consumption, still less for profligate waste. Nature has provided against the absolute destruction of any of her elementary matter, the
raw material of her works; the thunderbolt and the tornado, the most con-
vulsive throes of even the volcano and the earthquake, being only phenomena of decomposition and recomposition. But she has left it within
the power of man irreparably to derange the combinations of inorganic
matter and of organic life, which through the night of æons she had been
proportioning and balancing, to prepare the earth for his habitation, when,
in the fulness of time, his Creator should call him forth to enter into its
possession.

Apart from the hostile influence of man, the organic and the inorganic
world are, as I have remarked, bound together by such mutual relations and
adaptations as secure, if not the absolute permanence and equilibrium of
both, a long continuance of the established conditions of each at any given
time and place, or at least, a very slow and gradual succession of changes in
those conditions. But man is everywhere a disturbing agent. Wherever he
plants his foot, the harmonies of nature are turned to discords. The proportions and accommodations which insured the stability of existing arrangements are overthrown. Indigenous vegetable and animal species are extirpated, and supplanted by others of foreign origin, spontaneous production
is forbidden or restricted, and the face of the earth is either laid bare or covered with a new and reluctant growth of vegetable forms, and with alien
tribes of animal life. These intentional changes and substitutions constitute, indeed, great revolutions; but vast as is their magnitude and importance, they are, as we shall see, insignificant in comparison with the contingent and unsought results which have flowed from them.

The fact that, of all organic beings, man alone is to be regarded as essentially a destructive power, and that he wields energies to resist which,
nature—that nature whom all material life and all inorganic substance
obey—is wholly impotent, tends to prove that, though living in physical
nature, he is not of her, that he is of more exalted parentage, and belongs to
a higher order of existence than those born of her womb and submissive to
her dictates.

There are, indeed, brute destroyers, beasts and birds and insects of prey—all animal life feeds upon, and, of course, destroys other life,—but this destruction is balanced by compensations. It is, in fact, the very means by which the existence of one tribe of animals or of vegetables is secured against being smothered by the encroachments of another; and the reproductive powers of species, which serve as the food of others, are always proportioned to the demand they are destined to supply. Man pursues his victims with reckless destructiveness; and, while the sacrifice of life by the lower animals is limited by the cravings of appetite, he unsparingly persecutes, even to extirpation, thousands of organic forms which he cannot consume.

(The terrible destructiveness of man is remarkably exemplified in the chase of large mammalia and birds for single products, attended with the entire waste of enormous quantities of flesh, and of other parts of the animal, which are capable of valuable uses. The wild cattle of South America are slaughtered by millions for their hides and horns; the buffalo of North America for his skin or his tongue; the elephant, the walrus, and the narwhal for their tusks, the cetacea, and some other marine animals, for their oil and whalebone; the ostrich and other large birds, for their plumage. Within a few years, sheep have been killed in New England by whole flocks, for their pelts and suet alone, the flesh being thrown away; and it is even said that the bodies of the same quadrupeds have been used in Australia as fuel for lime kilns. What a vast amount of human nutriment, of bone, and of other animal products valuable in the arts, is thus recklessly squandered! In nearly all these cases, the part which constitutes the motive for this wholesale destruction, and is alone saved, is essentially of insignificant value as compared with what is thrown away. The horns and hide of an ox are not economically worth a tenth part as much as the entire carcass. One of the greatest benefits to be expected from the improvements of civilization is, that increased facilities of communication will render it possible to transport to places of consumption much valuable material that is now wasted because the price at the nearest market will not pay freight. The cattle slaughtered in South America for their hides would feed millions of the starving population of the Old World, if their flesh could be economically preserved and transported across the ocean.)

The earth was not, in its natural condition, completely adapted to the use of man, but only to the sustenance of wild animals and wild vegetation. These live, multiply their kind in just proportion, and attain their perfect measure of strength and beauty, without producing or requiring any change in the natural arrangements of surface, or in each other's spontaneous tendencies, except such mutual repression of excessive increase as may prevent the extirpation of one species by the encroachments of another. In short, without man, lower animal and spontaneous vegetable life would have been constant in type, distribution, and proportion, and the physical geography of the earth would have remained undisturbed for indefinite periods, and been subject to revolution only from possible, unknown cosmical causes, or from geological action.

But man, the domestic animals that serve him, the field and garden plants the products of which supply him with food and clothing, cannot subsist and rise to the full development of their higher properties, unless brute and unconscious nature be effectually combated, and, in a great degree, vanquished by human art. Hence, a certain measure of transformation of terrestrial surface, of suppression of natural, and stimulation of artificially modified productivity becomes necessary. This measure man has unfortunately exceeded. He has felled the forests whose network of fibrous roots bound the mould to the rocky skeleton of the earth; but had he allowed here and there a belt of woodland to reproduce itself by spontaneous propagation, most of the mischiefs which his reckless destruction of the natural protection of the soil has occasioned would have been averted. He has broken up the mountain reservoirs, the percolation of whose waters through unseen channels supplied the fountains that refreshed his cattle and fertilized his fields; but he has neglected to maintain the cisterns and the canals of irrigation which a wise antiquity had constructed to neutralized the consequences of its own imprudence. While he has torn the thin glebe which confined the light earth of extensive plains, and has destroyed the fringe of semi-aquatic plants which skirted the coast and checked the drifting of the sea sand, he has failed to prevent the spreading of the dunes by clothing them with artificially propagated vegetation. He has ruthlessly warred on all the tribes of animated nature whose spoil he could convert to his own uses, and he has not protected the birds which prey on the insects most destructive to his own harvests.

Purely untutored humanity, it is true, interferes comparatively little with the arrangements of nature, and the destructive agency of man becomes more and more energetic and unsparing as he advances in civilization, until the impoverishment, with which his exhaustion of the natural resources of the soil is threatening him, at last awakens him to the necessity of preserving what is left, if not of restoring what has been wantonly wasted. The wandering savage grows no cultivated vegetable, fells no forest, and extirpates no useful plant, no noxious weed. If his skill in the chase enables him to entrap numbers of the animals on which he feeds, he compensates this loss by destroying also the lion, the tiger, the wolf, the otter, the seal, and the eagle, thus indirectly protecting the feebler quadrupeds and fish and fowls, which would otherwise become the booty of beasts and birds of prey. But with stationary life, or rather with the pastoral state, man at once commences an almost indiscriminate warfare upon all the forms of animal and vegetable existence around him, and as he advances in civilization, he gradually eradicates or transforms every spontaneous product of the soil he occupies.

Human and Brute Action Compared

It has been maintained by authorities as high as any known to modern science, that the action of man upon nature, though greater in degree, does not differ in kind, from that of wild animals. It appears to me to differ in essential character, because, though it is often followed by unforeseen and undesired results, yet it is nevertheless guided by a self-conscious and intelligent will aiming as often at secondary and remote as at immediate objects. The wild animal, on the other hand, acts instinctively, and, so far as we are able to perceive, always with a view to single and direct purposes. The backwoodsman and the beaver alike fell trees; the man that he may convert the forest into an olive grove that will mature its fruit only for a succeeding generation, the beaver that he may feed upon their bark or use them in the construction of his habitation. Human differs from brute action, too, in its influence upon the material world, because it is not controlled by natural compensations and balances. Natural arrangements, once disturbed by man, are not restored until he retires from the field, and leaves free scope to spontaneous recuperative energies; the wounds he inflicts upon the material creation are not healed until he withdraws the arm that gave the blow. On the other hand, I am not aware of any evidence that wild animals have ever destroyed the smallest forest, extirpated any organic species, or modified its natural character, occasioned any permanent change of terrestrial surface, or produced any disturbance of physical conditions which nature has not, of herself, repaired without the expulsion of the animal that had caused it.

The form of geographical surface, and very probably the climate of a given country, depend much on the character of the vegetable life belonging to it. Man has, by domestication, greatly changed the habits and properties of the plants he rears; he has, by voluntary selection, immensely modified the forms and qualities of the animated creatures that serve him; and he has, at the same time, completely rooted out many forms of both vegetable and animal being. What is there, in the influence of brute life, that corresponds to this? We have no reason to believe that in that portion of the American continent which, though peopled by many tribes of quadruped and fowl, remained uninhabited by man, or only thinly occupied by purely savage tribes, any sensible geographical change had occurred within twenty centuries before the epoch of discovery and colonization, while, during the same period, man had changed millions of square miles, in the fairest and most fertile regions of the Old World, into the barrenest deserts.

The ravages committed by man subvert the relations and destroy the balance which nature had established between her organized and her inorganic creations; and she avenges herself upon the intruder, by letting loose upon her defaced provinces destructive energies hitherto kept in check by organic forces destined to be his best auxiliaries, but which he has unwisely dispersed and driven from the field of action. When the forest is gone, the great reservoir of moisture stored up in its vegetable mould is evaporated, and returns only in deluges of rain to wash away the parched dust into which that mould has been converted. The well-wooded and humid hills are turned to ridges of dry rock, which encumbers the low grounds and chokes the watercourses with its debris, and—except in countries favored with an equable distribution of rain through the seasons, and a moderate and regular inclination of surface—the whole earth, unless rescued by human art from the physical degradation to which it tends, becomes an assemblage of bald mountains, of barren, turfless hills, and of swampy and malarious plains. There are parts of Asia Minor, of Northern Africa, of Greece, and even of Alpine Europe, where the operation of causes set in action by man has brought the face of the earth to a desolation almost as complete as that of the moon; and though, within that brief space of time which we call "the historical period," they are known to have been covered with luxuriant woods, verdant pastures, and fertile meadows, they are now too far deteriorated to be reclaimable by man, nor can they become again fitted for human use, except through great geological changes, or other mysterious influences or agencies of which we have no present knowledge, and over which have no prospective control. The earth is fast becoming an unfit home for its noblest inhabitant, and another era of equal human crime and human improvidence, and of like duration with that through which traces of that crime and that improvidence extend, would reduce it to such a condition of impoverished productiveness, of shattered surface, of climatic excess, as to threaten the depravation, barbarism, and perhaps even extinction of the species.

Physical Improvement

True, there is a partial reverse to this picture. On narrow theatres, new forests have been planted; inundations of flowing streams restrained by heavy walls of masonry and other constructions; torrents compelled to aid, by depositing the slime with which they are charged, in filling up lowland, and raising the level of morasses which their own overflows had created; ground submerged by the encroachments of the ocean, or exposed to be covered by its tides, has been rescued from its dominion by diking; swamps and even lakes have been drained, and their beds brought within the domain of agricultural industry; drifting coast dunes have been checked and made productive by plantation; seas and inland waters have been repeopled with fish, and even the sands of the Sahara have been fertilized by ascertain fountains. These achievements are more glorious than the proudest triumphs of war, but, thus far, they give but faint hope that we shall yet make full atonement for our spendthrift waste of the bounties of nature.

Arrest of Physical Decay of New Countries

Comparatively short as is the period through which the colonization of foreign lands by European emigrants extends, great, and, it is to be feared, sometimes irreparable, injury has been already done in the various processes by which man seeks to subjugate the virgin earth; and many provinces, first trodden by the *Homo sapiens Europæ* within the last two centuries, begin to show signs of that melancholy dilapidation which is now driving so many of the peasantry of Europe from their native hearths. It is evidently a matter of great moment, not only to the population of the states where these symptoms are manifesting themselves, but to the general interests of humanity, that this decay should be arrested, and that the future operations of rural husbandry and of forest industry, in districts yet remaining substantially in their native condition, should be so conducted as to prevent the widespread mischiefs which have been elsewhere produced by thoughtless or wanton destruction of the natural safeguards of the soil. This can be done only by the diffusion of knowledge on this subject among the classes that, in earlier days, subdued and tilled ground in which they had no vested rights, but who, in our time, own their woods, their pastures, and their ploughlands as a perpetual possession for them and theirs, and have, therefore, a strong interest in the protection of their domain against deterioration.

Chapter II: Transfer, Modification, and Extirpation of Vegetable and of Animal Species

Marsh devoted most of *Man and Nature* to discussions of transformations of entire landscapes, arguing that landscapes transformed by humans do not function as well as natural landscapes, even if the transformation is one of degree (e.g., forests to fields) rather than of kind (e.g., forests to cities). He noted this phenomenon as an argument for why humans need to be considered a transformative force on Earth.

Yet in this chapter Marsh focused largely on the impacts of humans on species, especially on our role in bringing about extinction and introducing exotic species. He described eloquently the phenomenon of exotic species and documented many examples of species introduction and spread. Sadly, the spread of exotics has dramatically increased since Marsh's time, fueled primarily by the globalization of human cultures and economies; more people and more goods are being transported more places more often than ever before. As a consequence, the introduction of exotics is considered to be the second greatest cause of extinction today, after habitat modification. The economic consequences of exotic species are almost incalculable. A recent study determined that a mere 79 exotic species have caused more than $97 billion of damage between 1900 and 1993 in the United States alone. In the context of the many thousands of species introduced around the world since the rise and expansion of global travel, the total economic impact of exotics is unimaginable. Once established, exotics can be eradicated, if at all, only by massive campaigns of biological and chemical control. Such campaigns are impractical in almost all cases; therefore, one of the unavoidable and unfortunate consequences of the globalization of human culture is a globalization and impoverishment of the biological world.

Marsh correctly noted that the introduction of exotics is brought about by more than one force. Some species are introduced on purpose. Marsh documented species introduced for economic value and biological control. Today we are also aware of species introduced for aesthetics and recreation. The greater percentage of introductions, however, come from accidental release.

Marsh seemed to hold simultaneously two contradictory views about exotic species. One the one hand, the introduction of species for agricultural or other economic reasons may "be employed with advantage." He promoted this view in earlier writings, such as "The Camel" (1855) and "The Artificial Propagation of Fish" (1857). On the other hand, he argued forcefully that exotic species had great destructive powers. Marsh is not alone in holding these contradictory views. Our society today, even with abundant evidence of the destructive powers and economic impacts of exotic invaders, continues to allow the importation of species whose powers of invasion are unknown.

Marsh was also concerned about the role of humans in causing extinction. The hallmark of the conservation crisis today is still the extinction of species, which is presently occurring at a rate scarcely surpassed at any other point in Earth's history. Marsh developed three themes that are still widely accepted today. First, extinction caused by humans, both by cultures of European descent and cultures native to other regions, has occurred widely across the globe. Further, extinctions have been brought about in two distinct ways: direct exploitation and habitat modification. Conservation biologists today still recognize these as potent forces of extinction, although now consider habitat modification to have the greater impact of the two, and generally add to the list the introduction of exotic species.

Second, depletion of a species may well bring with it a cascade of ecological effects as other species respond to the release from predation or competition. Marsh discussed changes in marine invertebrates as a result of the depletion of whales. More recently we have come to recognize that the destruction of kelp forests in the Pacific Northwest has been caused by population explosions in sea urchins, grazers on kelp that were formerly kept in check by sea otters, which were hunted to near extinction up through the mid-1900s for their pelts.

Third, the cascade of ecological effects that come from the depletion of a species may, in fact, have serious negative consequences for human societies. One of the most important developments in conservation in recent years has been the widespread appreciation of this theme, along with efforts to critically assess the economic value of biodiversity and ecosystem functions.

Marsh also articulated the view that the human species itself could go extinct, and could avoid doing so only by paying closer attention to "the ways of nature."

Modern Geography embraces Organic Life

Whenever man has transported a plant from its native habitat to a new soil, he has introduced a new geographical force to act upon it, and this generally at the expense of some indigenous growth which the foreign vegetable has supplanted. The new and the old plants are rarely the equivalents of each other, and the substitution of an exotic for a native tree, shrub, or grass, increases or diminishes the relative importance of the vegetable element in the geography of the country to which it is removed. Further, man sows that he may reap. The products of agricultural industry are not suffered to rot upon the ground, and thus raise it by an annual stratum of new mould. They are gathered, transported to greater or less distances, and after they have served their uses in human economy, they enter, on the final decomposition of their elements, into new combinations, and are only in small proportion returned to the soil on which they grew. The roots of the grasses, and of many other cultivated plants, however, usually remain and decay in the earth, and contribute to raise its surface, though certainly not in the same degree as the forest.

The vegetables, which have taken the place of trees, unquestionably perform many of the same functions. They radiate heat, they condense the humidity of the atmosphere, they act upon the chemical constitution of the air, their roots penetrate the earth to greater depths than is commonly supposed, and form an inextricable labyrinth of filaments which bind the soil together and prevent its erosion by water. The broad-leaved annuals and perennials, too, shade the ground, and prevent the evaporation of moisture from its surface by wind and sun. At a certain stage of growth, grass land is probably a more energetic radiator and condenser than even the forest, but this powerful action is exerted, in its full intensity, for a few days only, while trees continue such functions, with unabated vigor, for many months in succession. Upon the whole, it seems quite certain, that no cultivated ground is as efficient in tempering climatic extremes, or in conservation of geographical surface and outline, as is the soil which nature herself has planted.

Modes of Introduction of Foreign Plants

. . . The seven hundred new species which have found their way to St. Helena within three centuries and a half, were certainly not all, or even in the largest proportion, designedly planted there by human art, and if we were well acquainted with vegetable emigration, we should probably be able to show that man has intentionally transferred fewer plants than he has accidentally introduced into countries foreign to them. After the wheat, follow the tares that infest it. The weeds that grow among the cereal grains, the pests of the kitchen garden, are the same in America as in Europe. The overturning of a wagon, or any of the thousand accidents which befall the emigrant in his journey across the Western plains, may scatter upon the ground the seeds he designed for his garden, and the herbs which fill so important a place in the rustic *materia medica* of the Eastern States, spring up along the prairie paths but just opened by the caravan of the settler. (Josselyn, who wrote about fifty years after the foundation of the first British colony in New England, says that the settlers at Plymouth had observed more than twenty English plants springing up spontaneously near their improvements. Every country has many plants not now, if ever, made use of by man, and therefore not designedly propagated by him, but which cluster around his dwelling, and continue to grow luxuriantly on the ruins of his rural habitation after he has abandoned it. The site of a cottage, the very foundation stones of which have been carried off, may often be recognized, years afterward, by the rank weeds which cover it, though no others of the same species are found for miles. . . .) The *hortus siccus* of a botanist may accidentally sow seeds from the foot of the Himalayas on the plains that skirt the Alps; and it is a fact of every familiar observation, that exotics, transplanted to foreign climates suited to their growth, often escape from the flower garden and naturalize themselves among the spontaneous vegetation of the pastures. When the cases containing the artistic treasures of Thorvaldsen were opened in the court of the museum where they are deposited, the straw and grass employed in packing them were scattered upon the ground, and the next season there sprang up from the seeds no less than twenty-five species of plants belonging to the Roman campagna, some of which were preserved and cultivated as a new tribute to the memory of the great Scandinavian sculptor, and at least four are said to have spontaneously naturalized themselves about Copenhagen. In the campaign of 1814, the Russian troops brought, in the stuffing of their saddles and by other accidental means, seeds from the banks of the Dnieper to the valley of the Rhine, and even introduced the plants of the steppes into the environs of Paris. The Turkish armies, in their incursions into Europe, brought Eastern vegetables in their train, and left the seeds of Oriental wall plants to grow upon the ramparts of Buda and Vienna. The Canada thistle, *Erigeron Canadense,* is said to have sprung up in Europe, two hundred years ago, from a seed which dropped out of the stuffed skin of a bird. (Accidents sometimes limit, as well as promote, the propagation of foreign vegetables in countries new to them. The Lombardy poplar is a diœcious tree, and is very easily grown from cuttings. In most of the countries into which it has been introduced the cuttings have been taken from the male, and as, consequently, males only have grown from them, the poplar does not produce seed in those regions. This is a fortunate circumstance, for otherwise this most worthless and least ornamental of trees would spread with a rapidity that would make it an annoyance to the agriculturist.)

Vegetables, how affected by Transfer to Foreign Soils

Vegetables, naturalized abroad either by accident or design, sometimes exhibit a greatly increased luxuriance of growth. The European cardoon, an esculent thistle, has broken out from the gardens of the Spanish colonies on the La Plata, acquired a gigantic stature, and propagated itself, in impenetrable thickets, over hundreds of leagues of the Pampas; and the *Anacharis alsinastrum*, a water plant not much inclined to spread in its native American habitat, has found its way into English rivers, and extended itself to such a degree as to form a serious obstruction to the flow of the current, and even to navigation.

Not only do many wild plants exhibit a remarkable facility of accommodation, but their seeds usually possess great tenacity of life, and their germinating power resists very severe trials. Hence, while the seeds of very many cultivated vegetables lose their vitality in two or three years, and can be transported safely to distant countries only with great precautions, the weeds that infest those vegetables, though not cared for by man, continue to accompany him in his migrations, and find a new home on every soil he colonizes. Nature fights in defence of her free children, but wars upon them when they have deserted her banners, and tamely submitted to the dominion of man. . . .

Extirpation of Quadrupeds

Although man never fails greatly to diminish, and is perhaps destined ultimately to exterminate, such of the larger wild quadrupeds as he cannot profitably domesticate, yet their numbers often fluctuate, and even after they seem almost extinct, they sometimes suddenly increase, without any intentional steps to promote such a result on his part. During the wars which followed the French Revolution, the wolf multiplied in many parts of Europe, partly because the hunters were withdrawn from the woods to chase a nobler game, and partly because the bodies of slain men and horses supplied this voracious quadruped with more abundant food. The same animal became again more numerous in Poland after the general disarming of the rural population by the Russian Government. On the other hand, when the hunters pursue the wolf, the graminivorous wild quadrupeds increase, and thus in turn promote the multiplication of their great four-footed destroyer by augmenting the supply of his nourishment. So long as the fur of the beaver was extensively employed as a material for fine hats, it bore a very high price, and the chase of this quadruped was so keen that naturalists feared its speedy extinction. When a Parisian manufacturer invented the silk hat, which soon came into almost universal use, the demand for beavers' fur fell off, and this animal—whose habits, as we have seen, are an important agency in the formation of bogs and other modifications of forest nature—immediately began to increase, reappeared in haunts which he had long abandoned, and can no longer be regarded as rare enough to be in immediate danger of extirpation. Thus the convenience or the caprice of Parisian fashion has unconsciously exercised an influence which may sensibly affect the physical geography of a distant continent.

Since the invention of gunpowder, some quadrupeds have completely disappeared from many European and Asiatic countries where they were formerly numerous. The last wolf was killed in Great Britain two hundred years ago, and the bear was extirpated from that island still earlier. The British wild ox exists only in a few English and Scottish parks, while in Irish bogs, of no great apparent antiquity, are found antlers which testify to the former existence of a stag much larger than any extant European species. The lion is believed to have inhabited Asia Minor and Syria, and probably Greece and Sicily also, long after the commencement of the historical period, and he is even said to have been not yet extinct in the first-named two of these countries at the time of the first Crusades. Two large graminivorous or browsing quadrupeds, the ur and the schelk, once common in Germany, are utterly extinct, the eland and the auerochs nearly so. . . . Modern naturalists identify the elk with the eland, the wisent with the auerochs. The period when the ur and the schelk became extinct is not known. The auerochs survived in Prussia until the middle of the last century, but unless it is identical with a similar quadruped said to be found on the Caucasus, it now exists only in the Russian imperial forest of Bialowitz, where about a thousand are still preserved, and in some great menageries, as for example that at Schönbrunn, near Vienna, which, in 1852, had four specimens. The eland, which is closely allied to the American wapiti, if not specifically the same animal, is still kept in the royal preserves of Prussia, to the number of four or five hundred individuals. The chamois is becoming rare, and the ibex or steinbock, once common in all the high Alps, is now believed to be confined to the Cogne mountains in Piemonte, between the valleys of the Dora Baltea and the Orco.

Birds as Sowers and Consumers of Seeds, and as Destroyers of Insects

. . . An unfortunate popular error greatly magnifies the injury done to the crops of grain and leguminous vegetables by wild birds. Very many of those generally supposed to consume large quantities of the seeds of cultivated plants really feed almost exclusively upon insects, and frequent the wheat-fields, not for the sake of the grain, but for the eggs, larvæ, and fly of the multiplied tribes of insect life which are so destructive to the harvests. This fact has been so well established by the examination of the stomachs of great numbers of birds in Europe and New England, at different seasons of the year, that it is no longer open to doubt, and it appears highly probable that even the species which consume more or less grain generally make amends, by destroying insects whose ravages would have been still more injurious. On this subject, we have much other evidence besides that derived from dis-section. Direct observation has shown, in many instances, that the destruc-tion of wild birds has been followed by a great multiplication of noxious in-sects, and, on the other hand; that these latter have been much reduced in numbers by the protection and increase of the birds that devour them.

Diminution and Extirpation of Birds

The general hostility of the European populace to the smaller birds is, in part, the remote effect of the reaction created by the game laws. When the restrictions imposed upon the chase by those laws were suddenly removed in France, the whole people at once commenced a destructive campaign against every species of wild animal. . . .

The French Revolution removed similar restrictions, with similar re-sults, in other countries. The habits then formed have become hereditary on the Continent, and though game laws still exist in England, there is lit-tle doubt that the blind prejudices of the ignorant and half-educated classes in that country against birds are, in some degree, at least, due to a legisla-tion, which, by restricting the chase of all game worth killing, drives the un-privileged sportsman to indemnify himself by slaughtering all wild life which is not reserved for the amusement of his betters. Hence the lord of the manor buys his partridges and his hares by sacrificing the bread of his tenants, and so long as the farmers of Crawley are forbidden to follow higher game, they will suicidally revenge themselves by destroying the sparrows which protect their wheatfields.

On the Continent, and especially in Italy, the comparative scarcity and dearness of animal food combine with the feeling I have just mentioned to stimulate still further the destructive passions of the fowler. In the Tuscan province of Grosseto, containing less than 2,000 square miles, nearly 300,000 thrushes and other small birds are annually brought to market.

Birds are less hardy in constitution, they possess less facility of accommodation, and they are more severely affected by climatic excess than quadrupeds. Besides, they generally want the means of shelter against the inclemency of the weather and against pursuit by their enemies, which holes and dens afford to burrowing animals and to some larger beasts of prey. The egg is exposed to many dangers before hatching, and the young bird is especially tender, defenceless, and helpless. Every cold rain, every violent wind, every hailstorm during the breeding season, destroys hundreds of nestlings, and the parent often perishes with her progeny while brooding over it in the vain effort to protect it. The great proportional numbers of birds, their migratory habits, and the ease with which they may escape most dangers that beset them, would seem to secure them from extirpation, and even from very great numerical reduction. But experience shows that when not protected by law, by popular favor or superstition, or by other special circumstances, they yield very readily to the hostile influences of civilization, and, though the first operations of the settler are favorable to the increase of many species, the great extension of rural and of mechanical industry is, in a variety of ways, destructive even to tribes not directly warred upon by man.

Nature sets bounds to the disproportionate increase of birds, while at the same time, by the multitude of their resources, she secures them from extinction through her own spontaneous agencies. Man both preys upon them and wantonly destroys them. The delicious flavor of game birds, and the skill implied in the various arts of the sportsman who devotes himself to fowling, make them favorite objects of the chase, while the beauty of their plumage, as a military and feminine decoration, threatens to involve the sacrifice of the last survivor of many once numerous species. Thus far, but few birds described by ancient or modern naturalists are known to have become absolutely extinct, though there are some cases in which they are ascertained to have utterly disappeared from the face of the earth in very recent times. The most familiar instances are those of the dodo, a large bird peculiar to the Mauritius or Isle of France, exterminated about the year 1690 and now known only by two or three fragments of skeletons, and the solitary, which inhabited the islands of Bourbon and Rodriguez, but has not been seen for more than a century. A parrot and some other birds of the Norfolk Island group are said to have lately become extinct. The wingless auk, *Alca impennis,* a bird remarkable for its excessive fatness, was very abundant two or three hundred years ago in the Faroe Islands, and on the whole Scandinavian seaboard. The early voyagers found either the same or a closely allied species, in immense numbers, on all the coasts and islands of Newfoundland. The value of its flesh and its oil made it one of the most important resources of the inhabitants of those sterile regions, and it was naturally an object of keen pursuit. It is supposed to be now completely extinct, and few museums can show even its skeleton.

There seems to be strong reason to believe that our boasted modern civilization is guiltless of one or two sins of extermination which have been committed in recent ages. New Zealand formerly possessed three species of dinornis, one of which, called *moa* by the islanders, was much larger than the ostrich. The condition in which the bones of these birds have been found and the traditions of the natives concur to prove that, though the aborigines had probably extirpated them before the discovery of New Zealand by the whites, they still existed at a comparatively late period. The same remarks apply to a winged giant the eggs of which have been brought from Madagascar. This bird must have much exceeded the dimensions of the moa, at least so far as we can judge from the egg, which is eight times as large as the average size of the ostrich egg, or about one hundred and fifty times that of the hen.

But though we have no evidence that man has exterminated many species of birds, we know that his persecutions have caused their disappearance from many localities where they once were common, and greatly diminished their numbers in others. The cappercailzie, *Tetrao urogallus*, the finest of the grouse family, formerly abundant in Scotland, had become extinct in Great Britain, but has been reintroduced from Sweden. The ostrich is mentioned by all the old travellers, as common on the Isthmus of Suez down to the middle of the seventeenth century. It appears to have frequented Syria and even Asia Minor at earlier periods, but is now found only in the seclusion of remoter deserts.

The modern increased facilities of transportation have brought distant markets within reach of the professional hunter, and thereby given a new impulse to his destructive propensities. Not only do all Great Britain and Ireland contribute to the supply of game for the British capital, but the canvas-back duck of the Potomac, and even the prairie hen from the basin of the Mississippi, may be found at the stalls of the London poulterer. Kohl informs us that on the coasts of the North Sea, twenty thousand wild ducks are usually taken in the coarse of the season in a single decoy, and sent to the large maritime towns for sale. The statistics of the great European cities show a prodigious consumption of game birds, but the official returns fall far below the truth, because they do not include the rural districts, and because neither the poacher nor his customers report the number of his victims. Reproduction, in cultivated countries, cannot keep pace with this excessive destruction, and there is no doubt that all the wild birds which are chased for their flesh or their plumage are diminishing with a rapidity which justifies the fear that the last of them will soon follow the dodo and the wingless auk.

Fortunately the larger birds which are pursued for their flesh or for their feathers, and those the eggs of which are used as food, are, so far as we know the functions appointed to them by nature, not otherwise specially useful to man, and, therefore, their wholesale destruction is an economical evil only in the same sense in which all waste of productive capital is an evil. If it were possible to confine the consumption of game fowl to a number equal to the annual increase, the world would be a gainer, but not to the same extent as it would be by checking the wanton sacrifice of millions of the smaller birds, which are of no real value as food, but which, as we have seen, render a most important service by battling, in our behalf, as well as in their own, against the countless legions of humming and of creeping things, with which the prolific powers of insect life would otherwise cover the earth.

Introduction of Birds

Man has undesignedly introduced into new districts perhaps fewer species of birds than of quadrupeds; but the distribution of birds is very much influenced by the character of his industry, and the transplantation of every object of agricultural production is, at a longer or shorter interval, followed by that of the birds which feed upon its seeds, or more frequently upon the insects it harbors. The vulture, the crow, and other winged scavengers, follow the march of the armies as regularly as the wolf. Birds accompany ships on long voyages, for the sake of the offal which is thrown overboard, and, in such cases, it might often happen that they would breed and become naturalized in countries where they had been unknown before. There is a familiar story of an English bird which built its nest in an unused block in the rigging of a ship, and made one or two short voyages with the vessel while hatching its eggs. Had the young become fledged while lying in a foreign harbor, they would of course have claimed the rights of citizenship in the country where they first took to the wing.

Introduction of Insects

The general tendency of man's encroachments upon spontaneous nature has been to increase insect life at the expense of vegetation and of the smaller quadrupeds and birds. Doubtless there are insects in all woods, but in temperate climates they are comparatively few and harmless, and the most numerous tribes which breed in the forest, or rather in its waters, and indeed in all solitudes, are those which little injure vegetation, such as mosquitoes, gnats, and the like. With the cultivated plants of man come the myriad tribes which feed or breed upon them, and agriculture not only introduces new species, but so multiplies the number of individuals as to defy calculation. Newly introduced vegetables frequently escape for years the insect plagues which had infested them in their native habitat; but the importation of other varieties of the plant, the exchange of seed, or some mere accident, is sure in the long run to carry the egg, the larvæ, or the chrysalis to the most distant shores where the plant assigned to it by nature as its possession has preceded it. For many years after the colonization of the United States, few or none of the insects which attack wheat in its different stages of growth, were known in America. During the Revolutionary war, the Hessian fly, *Cecidomyia destructor,* made its appearance, and it was so called because it was first observed in the year when the Hessian troops were brought over, and was popularly supposed to have been accidentally imported by those unwelcome strangers. Other destroyers of cereal grains have since found their way across the Atlantic, and a noxious European aphis has first attacked the American wheatfields within the last four or five years. Unhappily, in these cases of migration, the natural corrective of excessive multiplication, the parasitic or voracious enemy of the noxious insect, does not always accompany the wanderings of its prey, and the bane long precedes the antidote. Hence, in the United States, the ravages of imported insects injurious to cultivated crops, not being checked by the counteracting influences which nature had provided to limit their devastations in the Old World, are much more destructive than in Europe. It is not known that the wheat midge is preyed upon in America by any other insect, and in seasons favorable to it, it multiplies to a degree which would prove almost fatal to the entire harvest, were it not that, in the great territorial extent of the United States, there is room for such differences of soil and climate as, in a given year, to present in one State all the conditions favorable to the increase of a particular insect, while in another, the natural influences are hostile to it. The only apparent remedy for this evil is, to balance the disproportionate development of noxious foreign species by bringing from their native country the tribes which prey upon them. This, it seems, has been attempted. The United States Census Report for 1860, p. 82, states that the New York Agricultural Society "has introduced into this country from abroad certain parasites which Providence has created to counteract the destructive powers of some of these depredators."

This is, however, not the only purpose for which man has designedly introduced foreign forms of insect life. The eggs of the silkworm are known to have been brought from the farther East to Europe in the sixth century, and new silk spinners which feed on the castor oil bean and the ailanthus, have recently been reared in France and in South America with promising success. The cochineal, long regularly bred in aboriginal America, has been transplanted to Spain, and both the kermes insect and the cantharides have been transferred to other climates than their own. The honey bee must be ranked next to the silkworm in economical importance. This useful creature was carried to the United States by European colonists, in the latter part of the seventeenth century; it did not cross the Mississippi till the close of the eighteenth, and it is only within the last five or six years that it has been transported to California, where it was previously unknown. The Italian stingless bee has very lately been introduced into the United States.

The insects and worms intentionally transplanted by man bear but small proportion to those accidentally introduced by him. Plants and animals often carry their parasites with them, and the traffic of commercial countries, which exchange their products with every zone and every stage of social existence, cannot fail to transfer in both directions the minute organisms that are, in one way or another, associated with almost every object important to the material interest of man.

The tenacity of life possessed by many insects, their prodigious fecundity, the length of time they often remain in the different phases of their existence, the security of the retreats into which their small dimensions enable them to retire, are all circumstances very favorable not only to the perpetuity of their species, but to their transportation to distant climates and their multiplication in their new homes. The teredo, so destructive to shipping, has been carried by the vessels whose wooden walls it mines to almost every part of the globe. The termite, or white ant, is said to have been brought to Rochefort by the commerce of that port a hundred years ago. This creature is more injurious to wooden structures and implements than any other known insect. It eats out almost the entire substance of the wood, leaving only thin partitions between the galleries it excavates in it; but as it never gnaws through the surface to the air, a stick of timber may be almost wholly consumed without showing any external sign of the damage it has sustained. The termite is found also in other parts of France, and particularly at Rochelle, where, thus far, as its ravages are confined to a single quarter of the city. A borer, of similar habits, is not uncommon in Italy, and you may see in that country, handsome chairs and other furniture which have been reduced by this insect to a framework of powder of post, covered, and apparently held together, by nothing but the varnish.

Destruction of Fish

The inhabitants of the waters seem comparatively secure from human pursuit or interference by the inaccessibility of their retreats, and by our ignorance of their habits—a natural result of the difficulty of observing the ways of creatures living in a medium in which we cannot exist. Human agency has, nevertheless, both directly and incidentally, produced great changes in the population of the sea, the lakes, and the rivers, and if the effects of such revolutions in aquatic life are apparently of small importance in general geography, they are still not wholly inappreciable. The great diminution in the abundance of the larger fish employed for food or pursued for products useful in the arts is familiar, and when we consider how the vegetable and animal life on which they feed must be affected by the reduction of their numbers, it is easy to see that their destruction may involve considerable modifications in many of the material arrangements of nature. The whale does not appear to have been an object of pursuit by the ancients, for any purpose, nor do we know when the whale fishery first commenced. (I use whale not in a technical sense, but as a generic term for all the large inhabitants of the sea popularly grouped under that name.) It was, however, very actively prosecuted in the Middle Ages, and the Biscayans seem to have been particularly successful in this as indeed in other branches of nautical industry. Five hundred years ago, whales abounded in every sea. They long since became so rare in the Mediterranean as not to afford encouragement for the fishery as a regular occupation; and the great demand for oil and whalebone for mechanical and manufacturing purposes, in the present century, has stimulated the pursuit of the "hugest of living creatures" to such activity, that he has now almost wholly disappeared from many favorite fishing grounds, and in others is greatly diminished in numbers.

What special functions, besides his uses to man, are assigned to the whale in the economy of nature, we do not know; but some considerations, suggested by the character of the food upon which certain species subsist, deserve to be specially noticed. None of the great mammals grouped under the general name of whale are rapacious. They all live upon small organisms, and the most numerous species feed almost wholly upon the soft gelatinous mollusks in which the sea abounds in all latitudes. We cannot calculate even approximately the number of the whales, or the quantity of organic nutriment consumed by an individual, and of course we can form no estimate of the total amount of animal matter withdrawn by them, in a given period, from the waters of the sea. It is certain, however, that it must have been enormous when they were more abundant, and that it is still very considerable. A very few years since, the United States had more than six hundred whaling ships constantly employed in the Pacific, and the product of the American whale fishery for the year ending June 1st, 1860, was seven millions and a half of dollars. The mere bulk of the whales destroyed in a single year by the American and the European vessels engaged in this fishery would form an island of no inconsiderable dimensions, and each one of those taken must have consumed, in the course of his growth, many times his own weight of mollusks. The destruction of the whales must have been followed by a proportional increase of the organisms they feed upon, and if we had the means of comparing the statistics of these humble forms of life, for even so short a period as that between the years 1760 and 1860, we should find a difference sufficient possibly, to suggest an explanation of some phenomena at present unaccounted for.

For instance, as I have observed in another work, the phosphorescence of the sea was unknown to ancient writers, or at least scarcely noticed by them, and even Homer—who, blind as tradition makes him when he composed his epics, had seen, and marked, in earlier life, all that the glorious nature of the Mediterranean and its coasts discloses to unscientific observation—nowhere alludes to this most beautiful and striking of maritime wonders. In the passage just referred to, I have endeavored to explain the silence of ancient writers with respect to this as well as other remarkable phenomena on psychological grounds; but is it not possible that, in modern times, the animalculæ which produce it may have immensely multiplied, from the destruction of their natural enemies by man, and hence that the gleam shot forth by their decomposition, or by their living processes, is both more frequent and more brilliant than in the days of classic antiquity?

Although the whale does not prey upon smaller creatures resembling himself in form and habits, yet true fishes are extremely voracious, and almost every tribe devours unsparingly the feebler species, and even the spawn and young of its own. The enormous destruction of the pike, the trout family, and other ravenous fish, as well as of the fishing birds, the seal, and the otter, by man, would naturally have occasioned a great increase in the weaker and more defenceless fish on which they feed, had he not been as hostile to them also as to their persecutors. We have little evidence that any fish employed as human food has naturally multiplied in modern times, while all the more valuable tribes have been immensely reduced in numbers. This reduction must have affected the more voracious species not used as food by man, and accordingly the shark, and other fish of similar habits, though not objects of systematic pursuit, are now comparatively rare in many waters where they formerly abounded. The result is, that man has greatly reduced the numbers of all larger marine animals, and consequently indirectly favored the multiplication of the smaller aquatic organisms which entered into their nutriment. This change in the relations of the organic and inorganic matter of the sea must have exercised an influence on the latter. What that influence has been, we cannot say, still less can we predict what it will be hereafter; but its action is not for that reason the less certain.

Introduction and Breeding of Fish

The introduction and successful breeding of fish in foreign species appears to have been long practised in China and was not unknown to the Greek and Romans. This art has been revived in modern times, but thus far without any important results, economical or physical, though there seems to be good reason to believe it may be employed with advantage on an extended scale. As in the case of plants, man has sometimes undesignedly introduced new species of aquatic animals into countries distant from their birthplace. The accidental escape of the Chinese goldfish from ponds where they were bred as a garden ornament, has peopled some European, and it is said American streams with this species. Canals of navigation and irrigation interchange the fish of lakes and rivers widely separated by natural barriers, as well as the plants which drop their seeds into the waters. The Erie Canal, as measured by its own channel, has a length of about three hundred and sixty miles, and it has ascending and descending locks in both directions. By this route, the fresh-water fish of the Hudson and the Upper Lakes, and some of the indigenous vegetables of these respective basins, have intermixed, and the fauna and flora of the two regions have now more species common to both than before the canal was opened. . . .

The intentional naturalization of foreign fish, as I have said, has not thus far yielded important fruits; but though this particular branch of what is called, not very happily, pisciculture, has not yet established its claims to the attention of the physical geographer or the political economist, the artificial breeding of domestic fish has already produced very valuable results, and is apparently destined to occupy an extremely conspicuous place in the history of man's efforts to compensate his prodigal waste of the gifts of nature. The restoration of the primitive abundance of salt and fresh water fish, is one of the greatest material benefits that, with our present physical resources, governments can hope to confer upon their subjects. The rivers, lakes, and seacoasts once restocked, and protected by law from exhaustion

by taking fish at improper seasons, by destructive methods, and in extravagant quantities, would continue indefinitely to furnish a very large supply
of most healthful food, which, unlike all domestic and agricultural products, would spontaneously renew itself and cost nothing but the taking.
There are many sterile or wornout soils in Europe so situated that they
might, at no very formidable cost, be converted into permanent lakes,
which would serve not only as reservoirs to retain the water of winter rains
and snow, and give it out in the dry season for irrigation, but as breeding
ponds for fish, and would thus, without further cost, yield a larger supply of
human food than can at present be obtained from them even at a great expenditure of capital and labor in agricultural operations. The additions
which might be made to the nutriment of the civilized world by a judicious
administration of the resources of the waters, would allow some restriction
of the amount of soil at present employed for agricultural purposes, and a
corresponding extension of the area of the forest, and would thus facilitate
a return to primitive geographical arrangements which it is important partially to restore.

Extirpation of Aquatic Animals

It does not seem probable that man, with all his rapacity and all his enginery, will succeed in totally extirpating any salt-water fish, but he has already exterminated at least one marine warm-blooded animal—Steller's
sea cow—and the walrus, the sea lion, and other large amphibia, as well as
the principal fishing quadrupeds, are in imminent danger of extinction.
Steller's sea cow, *Rhytina Stelleri*, was first seen by Europeans in the year
1741, on Bering's Island. It was a huge amphibious mammal, weighing not
less than eight thousand pounds, and appears to have been confined exclusively to the islands and coasts in the neighborhood of Bering's Strait. Its
flesh was very palatable, and the localities it frequented were easily accessible from the Russian establishments in Kamtschatka. As soon as its existence and character, and the abundance of fur animals in the same waters,
were made known to the occupants of those posts by the return of the sur-
vivors of Bering's expedition, so active a chase was commenced against the
amphibia of that region, that, in the course of twenty-seven years, the sea
cow, described by Steller as extremely numerous in 1741, is believed to have
been completely extirpated, not a single individual having been seen since
the year 1768. The various tribes of seals in the Northern and Southern
Pacific, the walrus and the sea otter, are already so reduced in numbers
that they seem destined soon to follow the sea cow, unless protected by

legislation stringent enough, and a police energetic enough, to repress the ardent cupidity of their pursuers. . . .

Man has promoted the multiplication of fish by making war on their brute enemies, but he has by no means thereby compensated his own greater destructiveness. The bird and beast of prey, whether on land or in the water, hunt only as long as they feel the stimulus of hunger, their ravages are limited by the demands of present appetite, and they do not wastefully destroy what they cannot consume. Man, on the contrary, angles today that he may dine to-morrow; he takes and dries millions of fish on the banks of Newfoundland, that the fervent Catholic of the shores of the Mediterranean may have wherewithal to satisfy the cravings of the stomach during next year's Lent, without imperilling his soul by violating the discipline of the papal church; and all the arrangements of his fisheries are so organized as to involve the destruction of many more fish than are secured for human use, and the loss of a large proportion of the annual harvest of the sea in the process of curing, or in transportation to the places of its consumption.

(The indiscriminate hostility of man to inferior forms of animated life is little creditable to modern civilization, and it is painful to reflect that it becomes keener and more unsparing in proportion to the refinement of the race. The savage slays no animal, not even the rattlesnake, wantonly; and the Turk, whom we call a barbarian, treats the dumb beast as gently as a child. One cannot live many weeks in Turkey without witnessing touching instances of the kindness of the people to the lower animals, and I have found it very difficult to induce even the boys to catch lizards and other reptiles for preservation as specimens. The fearless confidence in man, so generally manifested by wild animals in newly discovered islands, ought to have inspired a gentler treatment of them; but a very few years of the relentless pursuit, to which they are immediately subjected, suffice to make them as timid as the wildest inhabitants of the European forest. This timidity, however, may easily be overcome. The squirrels introduced by Mayor Smith into the public parks of Boston are so tame as to feed from the hands of passengers, and they not unfrequently enter the neighboring houses.) . . .

Man has hitherto hardly anywhere produced such climatic or other changes as would suffice of themselves totally to banish the wild inhabitants of the dry land, and the disappearance of the native birds and quadrupeds from particular localities is to be ascribed quite as much to his direct persecutions as to the want of forest shelter, of appropriate food, or of other conditions indispensable to their existence. But almost all the processes of agriculture, and of mechanical and chemical industry, are fatally destructive to aquatic animals within reach of their influence. When, in consequence of clearing the woods, the changes already described as thereby produced in the beds and currents of rivers, are in progress, the spawning grounds of fish are exposed from year to year to a succession of mechanical disturbances; the temperature of the water is higher in summer, colder in winter, than when it was shaded and protected by wood; the smaller organisms, which formed the sustenance of the young fry, disappear or are reduced in numbers, and new enemies are added to the old foes that preyed upon them; the increased turbidness of the water in the annual inundations chokes the fish; and, finally, the quickened velocity of its current sweeps them down into the larger rivers or into the sea, before they are yet strong enough to support so great a change of circumstances. Industrial operations are not less destructive to fish which live or spawn in fresh water. Milldams impede their migrations, if they do not absolutely prevent them, the sawdust from lumber mills clogs their gills, and the thousand deleterious mineral substances, discharged into rivers from metallurgical, chemical, and manufacturing establishments, poison them by shoals.

Minute Organisms

. . . Nature has no unit of magnitude by which she measures her works. Man takes his standards of dimension from himself. The hair's breadth was his minimum until the microscope told him that there are animated creatures to which one of the hairs of his head is a larger cylinder than is the trunk of the giant California redwood to him. He borrows his inch from the breadth of his thumb, his palm and span from the width of his hand or the spread of his fingers, his foot from the length of the organ so named; his cubit is the distance from the tip of his middle finger to his elbow, and his fathom is the space he can measure with his outstretched arms. To a being who instinctively finds the standard of all magnitudes in his own material frame, all objects exceeding his own dimensions are absolutely great, all falling short of them absolutely small. Hence we habitually regard the whale and the elephant as essentially large and therefore important creatures, the animalcule as an essentially small and therefore unimportant organism. But no geological formation owes its origin to the labors or the remains of the huge mammal, while the animalcule composes, or has furnished, the substance of strata thousands of feet in thickness and extending, in unbroken beds, over many degrees of terrestrial surface. If man is destined to inhabit the earth much longer, and to advance in natural knowledge with the rapidity which has marked his progress in physical science for the last two or three centuries he will learn to put a wiser estimate on the works of creation, and will derive not only great instruction from studying the ways of nature in her obscurest, humblest walks, but great material advantage from stimulating her productive energies in provinces of her empire hitherto regarded as forever inaccessible, utterly barren.

Chapter III: The Woods

In this chapter, Marsh addressed conservation issues associated with forests. His thoughts here arguably represent his most creative and synthetic thinking, perhaps because he witnessed directly the impacts of deforestation as a child in Vermont. This visceral understanding of forest clearing, coupled with an intellectual appreciation for both the biology and geology of forestlands, derived from his readings, led Marsh to recognize the complex connections between forests, climate, soil, and water, connections that further research since Marsh's time has validated and clarified.

Among the points Marsh makes in this chapter that remain central tenets in conservation biology are:

• The removal of forest trees leads to flooding, landslides, erosion, losses of understory and ground-cover plants, increases in economically harmful insects, and decreases in water supply. Sadly, we have not yet fully learned to appreciate some of these connections: clear-cut logging in the Pacific Northwest even today causes dramatic and lethal landslides as bare slopes break away under the weight of accumulated soil moisture from rain.

• Domestic herbivores, such as cattle and sheep, have different effects on natural plant communities than do wild herbivores like deer and moose; therefore, domestic animals are not ecological substitutes for wild animals.

• The effects of deforestation may take time to be felt. The slow response time of ecological systems to disturbance is today used as a powerful argument for caution in the management of human-modified natural communities.

• The average sizes of trees in forests have declined as a result of the intensity and frequency of cutting. This trend has continued to the present day.

• Forests can and should be restored, a conservation strategy that requires public land.

• Forest harvesting practices should be designed to promote long-term production yields of forest products, a goal that today is referred to as sustainable forestry.

• Forest structure, especially the condition of the ground cover, affects climate. Although what Marsh understood about the details of this relationship was

limited by the scientific understanding of the nineteenth century, his observations of this relationship were of the type that paved the way for climate modeling a hundred years later and our understanding of the connection between
land-use practices and local climate change.

• With respect to conservation, the "government" is neither inherently good or
bad. Its structure and policies can both create environmental problems and lead
to the ability to recognize and solve them.

• Understory and ground-cover plants have their own economic values, including "valuable medicinal properties." Although Marsh discounted this as a likely
compelling argument for conservation, the role of forest plants and animals as
sources for medicines is today an important economic argument used in international treaty negotiations for the conservation of tropical forests.

• The people alive in any one generation have a responsibility to future generations to manage forests wisely.

Influence of the Forest, Considered as Inorganic Matter, on Temperature: a. Absorbing and Emitting Surface

A given area of ground, as estimated by the every-day rule of measurement in yards or acres, presents always the same apparent
quantity of absorbing, radiating, and reflecting surface; but the real extent
of that surface is very variable, depending, as it does, upon its configuration,
and the bulk and form of the adventitious objects it bears upon it; and,
besides, the true superficies remaining the same, its power of absorption,
radiation, reflection, and conduction of heat will be much affected by its
consistence, its greater or less humidity, and its color, as well as by its inclination of plane and exposure....

... If we suppose forty trees to be planted on an acre, one being situated
in the centre of every square of two rods the side [*1089 square feet*], and to
grow until their branches and leaves everywhere meet, it is evident that,
when in full foliage, the trunks, branches, and leaves would present an
amount of thermoscopic surface much greater than that of an acre of bare
earth; and besides this, the fallen trees lying scattered on the ground,
would somewhat augment the sum total....

It must further be remembered that the form and texture of a given surface are important elements in determining its thermoscopic character. Leaves are porous, and admit air and light more or less freely into their substance; they are generally smooth and even glazed on one surface; they are usually covered on one or both sides with spiculæ, and they very commonly present one or more acuminated points in their outline—all circumstances which tend to augment their power of emitting heat by reflection or radiation. Direct experiment on growing trees is very difficult, nor is it in any case practicable to distinguish how far a reduction of temperature produced by vegetation is due to radiation, and how far to exhalation of the fluids of the plant in a gaseous form; for both processes usually go on together. But the frigorific effect of leafy structure is well observed in the deposit of dew and the occurrence of hoarfrost on the foliage of grasses, and other small vegetables, and on other objects of similar form and consistence, when the temperature of the air a few yards above has not been brought down to the dew point, still less to 32°, the degree of cold required to congeal dew to frost.

e. Trees as a Shelter to Ground to the Leeward

The action of the forest, considered merely as a mechanical shelter to grounds lying to the leeward of it, would seem to be an influence of too restricted a character to deserve much notice; but many facts concur to show that it is an important element in local climate, and that it is often a valuable means of defence against the spread of miasmatic effluvia, though, in this last case, it may exercise a chemical as well as a mechanical agency. In the report of a committee appointed in 1836 to examine an article of the forest code of France, Arago observes: "If a curtain of forest on the coasts of Normandy and of Brittany were destroyed, these two provinces would become accessible to the winds from the west, to the mild breezes of the sea. Hence a decrease of the cold of winter. If a similar forest were to be cleared on the eastern border of France, the glacial east wind would prevail with greater strength, and the winters would become more severe. Thus the removal of a belt of wood would produce opposite effects in the two regions." . . .

The felling of the woods on the Atlantic coast of Jutland has exposed the soil not only to drifting sands, but to sharp sea winds, that have exerted a sensible deteriorating effect on the climate of the peninsula, which has no mountains to serve at once as a barrier to the force of the winds, and as a storehouse of moisture received by precipitation or condensed from atmospheric vapors.

It is evident that the effect of the forest, as a mechanical impediment to the passage of the wind, would extend to a very considerable distance above its own height, and hence protect while standing, or lay open when felled, a much larger surface than might at first thought be supposed. The atmosphere, movable as are its particles, and light and elastic as are its masses, is nevertheless held together as a continuous whole by the law of attraction between its atoms, and, therefore, an obstruction which mechanically impedes the movement of a given stratum of air, will retard the passage of the strata above and below it. To this effect may often be added that of an ascending current from the forest itself, which must always exist when the atmosphere within the wood is warmer than the stratum of air above it, and must be of almost constant occurrence in the case of cold winds, from whatever quarter, because the still air in the forest is slow in taking up the temperature of the moving columns and currents around and above it. Experience, in fact, has shown that mere rows of trees, and even much lower obstructions, are of essential service in defending vegetation against the action of the wind. . . .

The local retardation of spring so much complained of in Italy, France, and Switzerland, and the increased frequency of late frosts at that season, appear to be ascribable to the admission of cold blasts to the surface, by the felling of the forests which formerly both screened it as by a wall, and communicated the warmth of their soil to the air and earth to the leeward. Caimi states that since the cutting down of the woods of the Apennines, the cold winds destroy or stunt the vegetation, and that, in consequence of "the usurpation of winter on the domain of spring," the district of Mugello has lost all its mulberries, except the few which find in the lee of buildings a protection like that once furnished by the forest. . . .

Dussard, as quoted by Ribbe, maintains that even the mistral, or northwest wind, whose chilling blasts are so fatal to tender vegetation in the spring, "is the child of man, the result of his devastations." "Under the reign of Augustus," continues he, "the forests which protected the Cévennes were felled, or destroyed by fire, in mass. A vast country, before covered with impenetrable woods—powerful obstacles to the movement and even to the formation of hurricanes—was suddenly denuded, swept bare, stripped, and soon after, a scourge hitherto unknown struck terror over the land from Avignon to the Bouches du Rhone, thence to Marseilles, and then extended its ravages, diminished indeed by a long career which had partially exhausted its force, over the whole maritime frontier. The people thought this wind a curse sent of God. They raised altars to it and offered sacrifices to appease its rage." It seems, however, that this plague was less destructive than at present, until the close of the sixteenth century, when further clearings had removed most of the remaining barriers to its course. Up to that time, the northwest wind appears not to have attained to the maximum of specific effect which now characterizes it as a local phenomenon. Extensive districts, from which the rigor of the seasons has now banished valuable crops, were not then exposed to the loss of their harvests by tempests, cold, or drought. The deterioration was rapid in its progress. Under the Consulate, the clearings had exerted so injurious an effect upon the climate, that the cultivation of the olive had retreated several leagues, and since the winters and springs of 1820 and 1836, this branch of rural industry has been abandoned in a great number of localities where it was advantageously pursued before. The orange now flourishes only at a few sheltered points of the coast, and it is threatened even at Hyères, where the clearing of the hills near the town has proved very prejudicial to this valuable tree.

Marschand informs us that, since the felling of the woods, late spring frosts are more frequent in many localities north of the Alps; that fruit trees thrive well no longer, and that it is difficult to raise young trees.

General Consequence of the Destruction of the Forest

With the disappearance of the forest, all is changed. At one season, the earth parts with its warmth by radiation to an open sky—receives, at another, an immoderate heat from the unobstructed rays of the sun. Hence the climate becomes excessive, and the soil is alternately parched by the fervors of summer, and scarred by the rigors of winter. Bleak winds sweep unresisted over its surface, drift away the snow that sheltered it from the frost, and dry up its scanty moisture. The precipitation becomes as regular as the temperature; the melting snows and vernal rains, no longer absorbed by a loose and bibulous vegetable mould, rush over the frozen surface, and pour down the valleys seaward, instead of filling a retentive bed of absorbent earth, and storing up a supply of moisture to feed perennial springs. The soil is bared of its covering of leaves, broken and loosened by the plough, deprived of the fibrous rootlets which held it together, dried and pulverized by sun and wind, and at last exhausted by new combinations. The face of the earth is no longer a sponge, but a dust heap, and the floods which the waters of the sky pour over it hurry swiftly along its slopes, carrying in suspension vast quantities of earthy particles which increase the abrading power and mechanical force of the current, and, augmented by the sand and gravel of falling banks, fill the beds of the streams, divert them into new channels and obstruct their outlets. The rivulets, wanting their former regularity of supply and deprived of the protecting shade of the woods, are heated, evaporated, and thus reduced in their summer currents, but swollen to raging torrents in autumn and in spring. From these causes, there is a constant degradation of the uplands, and a consequent elevation of the beds of watercourses and of lakes by the deposition of the mineral and vegetable matter carried down by the waters. The channels of great rivers become unnavigable, their estuaries are choked up, and harbors which once sheltered large navies are shoaled by dangerous sandbars. The earth, stripped of its vegetable glebe, grows less and less productive, and, consequently, less able to protect itself by weaving a new network of roots to bind its particles together, a new carpeting of turf to shield it from wind and sun and scouring rain. Gradually it becomes altogether barren. The washing of the soil from the mountains leaves bare ridges of sterile rock, and the rich organic mould which covered them, now swept down into the dank low grounds, promotes a luxuriance of aquatic vegetation that breeds fever, and more insidious forms of mortal disease, by its decay, and thus the earth is rendered no longer fit for the habitation of man.

To the general truth of this sad picture there are many exceptions, even in countries of excessive climates. Some of these are due to favorable conditions of surface, of geological structure, and of the distribution of rain; in many others, the evil consequences of man's improvidence have not yet been experienced, only because a sufficient time has not elapsed, since the felling of the forest, to allow them to develop themselves. But the vengeance of nature for the violation of her harmonies, though slow, is sure, and the gradual deterioration of soil and climate in such exceptional regions is as certain to result from the destruction of the woods as is any natural effect to follow its cause.

In the vast farrago of crudities which the elder Pliny's ambition of encyclopædic attainment and his ready credulity have gathered together, we meet some judicious observations. Among these we must reckon the remark with which he accompanies his extraordinary statement respecting the prevention of springs by the growth of forest trees, though, as is usual with him, his philosophy is wrong. "Destructive torrents are generally formed when hills are stripped of the trees which formerly confined and absorbed the rains." The absorption here referred to is not that of the soil, but of the roots, which, Pliny supposed, drank up the water to feed the growth of the trees.

Although this particular evil effect of too extensive clearing was so early noticed, the lesson seems to have been soon forgotten. The legislation of the Middle Ages in Europe is full of absurd provisions concerning the forests, which sovereigns sometimes destroyed because they furnished a retreat for rebels and robbers, sometimes protected because they were necessary to breed stags and boars for the chase, and sometimes spared with the more enlightened view of securing a supply of timber and of fuel to future generations. It was reserved to later ages to appreciate their geographical importance, and it is only in very recent times, only in a few European countries, that too general felling of the woods has been recognized as the most destructive among the many causes of the physical deterioration of the earth.

The Influence of the Forest on Inundations

Besides the climatic question, which I have already sufficiently discussed, and the obvious inconveniences of a scanty supply of charcoal, of fuel, and of timber for architectural and naval construction, and for the thousand other uses to which wood is applied in rural and domestic economy, and in the various industrial processes of civilized life, the attention of French foresters and public economists has been specially drawn to three points, namely, the influence of the forests on the permanence and regular flow of springs or natural fountains; on inundations by the overflow of rivers; and on the abrasion of soil and the transportation of earth, gravel, pebbles, and even of considerable masses of rock, from higher to lower levels, by torrents. There are, however, connected with this general subject, several other topics of minor or strictly local interest, or of more uncertain character, which I shall have occasion more fully to speak of hereafter.

The first of these three principal subjects—the influence of the woods on springs and other living waters—has been already considered; and if the facts stated in the discussion are well established, and the conclusions I have drawn from them are logically sound, it would seem to follow, as a necessary corollary, that the action of the forest is as important in diminishing the frequency and violence of river floods, as in securing the permanence and equability of natural fountains; for any cause which promotes the absorption and accumulation of the water of precipitation by the superficial strata of the soil, to be slowly given out by infiltration and percolation, must, by preventing the rapid flow of surface water into the natural channels of drainage, tend to check the sudden rise of rivers, and, consequently, the overflow of their banks, which constitutes what is called inundation. The mechanical resistance, too, offered by the trunks of trees and of undergrowth to the flow of water over the surface, tends sensibly to retard the rapidity of its descent down declivities, and to divert and divide streams which may have already accumulated from smaller threads of water.

Inundations are produced by the insufficiency of the natural channels of rivers to carry off the waters of their basins as fast as those waters flow into them. In accordance with the usual economy of nature, we should presume that she had everywhere provided the means of discharging, without disturbance of her general arrangements or abnormal destruction of her products, the precipitation which she sheds upon the face of the earth. Observation confirms this presumption, at least in the countries to which I confine my inquiries; for, so far as we know the primitive conditions of the regions brought under human occupation within the historical period, it appears that the overflow of river banks was much less frequent and destructive than at the present day, or, at least, the rivers rose and fell less suddenly before man had removed the natural checks to the too rapid drainage of the basins in which their tributaries originate. The banks of the rivers and smaller streams in the North American colonies were formerly little abraded by the currents. Even now the trees come down almost to the water's edge along the rivers, in the larger forests of the United States, and the surface of the streams seems liable to no great change in level or in rapidity of current. A circumstance almost conclusive as to the regularity of flow in forest rivers, is that they do not form large sedimentary deposits, at their points of discharge into lakes or larger streams, such accumulations beginning, or at least advancing far more rapidly, after the valleys are cleared.

In the Northern United States, although inundations are sometimes produced in the height of summer by heavy rains it will be found generally true that the most rapid rise of the waters, and of course, the most destructive "freshets," as they are called in America, are produced by the sudden dissolution of the snow before the open ground is thawed in the spring. It frequently happens that a powerful thaw sets in after a long period of frost, and the snow which had been months in accumulating is dissolved and carried off in a few hours. When the snow is deep, it, to use a popular expression, "takes the frost out of the ground" in the woods, and, if it lies long enough, in the fields also. But the heaviest snows usually fall after midwinter, and are succeeded by warm rains or sunshine, which dissolve the snow on the cleared land before it has had time to act upon the frost-bound soil beneath it. In this case, the snow in the woods is absorbed as fast as it melts, by the soil it has protected from freezing and does not materially contribute to swell the current of the rivers. If the mild weather, in which great snowstorms usually occur, does not continue and become a regular thaw, it is almost sure to be followed by drifting winds, and the inequality with which they distribute the snow leaves the ridges comparatively bare, while the depressions are often filled with drifts to the height of many feet. The knolls become frozen to a great depth; succeeding partial thaws melt the surface snow, and the water runs down into the furrows of ploughed fields, and other artificial and natural hollows, and then often freezes to solid ice. In this state of things, almost the entire surface of the cleared land is impervious to water, and from the absence of trees and the general smoothness of the ground, it offers little mechanical resistance to superficial currents. If, under these circumstances, warm weather accompanied by rain occurs, the rain and melted snow are swiftly hurried to the bottom of the valleys and gathered to raging torrents.

It ought further to be considered that, though the lighter ploughed soils readily imbibe a great deal of water, yet the grass lands, and all the heavy and tenacious earths, absorb it in much smaller quantities, and less rapidly than the vegetable mould of the forest. Pasture meadow, and clayey soils, taken together, greatly predominate over the sandy ploughed fields, in all large agricultural districts, and hence, even if, in the case we are supposing, the open ground chance to have been thawed before the melting of the snow which covers it, it is already saturated with moisture, or very soon becomes so, and, of course, cannot relieve the pressure by absorbing more water. The consequence is that the face of the country is suddenly flooded with a quantity of melted snow and rain equivalent to a fall of six or eight inches of the latter, or even more. This runs unobstructed to rivers often still bound with thick ice, and thus inundations of a fearfully devastating character are produced. The ice bursts, from the hydrostatic pressure from below, or is violently torn up by the current, and swept by the impetuous stream, in large masses and with resistless fury, against banks, bridges, dams, and mills erected near them. The bark of the trees along the rivers is often abraded, at a height of many feet above the ordinary water level, by cakes of floating ice, which are at last stranded by the receding flood on meadow or ploughland, to delay, by their chilling influence, the advance of the tardy spring.

The surface of a forest, in its natural condition, can never pour forth such deluges of water as flow from cultivated soil. Humus, or vegetable mould, is capable of absorbing almost twice its own weight of water. The soil in a forest of deciduous foliage is composed of humus, more or less unmixed, to the depth of several inches, sometimes even of feet, and this stratum is usually able to imbibe all the water possibly resulting from the snow which at any one time covers it. But the vegetable mould does not cease to absorb water when it becomes saturated, for it then gives off a portion of its moisture to the mineral earth below, and thus is ready to receive a new supply; and, besides, the bed of leaves not yet converted to mould takes up and retains a very considerable proportion of snow water, as well as of rain.

In the warm climates of Southern Europe, as I have already said, the functions of forest, so far as the disposal of the water of precipitation is concerned, are essentially the same at all seasons, and are analogous to those which it performs in the Northern United States in summer. Hence, in the former countries, the winter floods have not the characteristics which mark them in the latter, nor is the conservative influence of the woods in winter relatively so important, though it is equally unquestionable.

If the summer floods in the United States are attended with less pecuniary damage than those of the Loire and other rivers of France, the Po and its tributaries in Italy, the Emme and her sister torrents which devastate the valleys of Switzerland, it is partly because the banks of American rivers are not yet lined with towns, their shores and the bottoms which skirt them not yet covered with improvements whose cost is counted by millions, and, consequently, a smaller amount of property is exposed to injury by inundation. But the comparative exemption of the American people from the terrible calamities which the overflow of rivers has brought on some of the fairest portions of the Old World, is, in a still greater degree, to be ascribed to the fact that, with all our thoughtless improvidence, we have not yet bared all the sources of our streams, not yet overthrown all the barriers which nature has erected to restrain her own destructive energies. Let us be wise in time, and profit by the errors of our older brethren! . . .

The destructive effects of inundations considered simply as a mechanical power by which life is endangered, crops destroyed, and the artificial constructions of man overthrown, are very terrible. Thus far, however, the flood is a temporary and by no means an irreparable evil, for if its ravages end here, the prolific powers of nature and the industry of man soon restore what had been lost, and the face of the earth no longer shows traces of the deluge that had overwhelmed it. Inundations have even their compensations. The structures they destroy are replaced by better and more secure erections, and if they sweep off a crop of corn, they not unfrequently leave behind them, as they subside, a fertilizing deposit which enriches the exhausted field for a succession of seasons. If, then, the too rapid flow of the surface waters occasioned no other evil than to produce, once in ten years upon the average, an inundation which should destroy the harvest of the low grounds along the rivers, the damage would be too inconsiderable, and of too transitory a character, to warrant the inconveniences and the expense involved in the measures which the most competent judges in many parts of Europe believe the respective governments ought to take to obviate it.

Destructive Action of Torrents

But the great, the irreparable, the appalling mischiefs which have already resulted, and threaten to ensue on a still more extensive scale hereafter,

from too rapid superficial drainage, are of a properly geographical character, and consist primarily in erosion, displacement, and transportation of
the superficial strata, vegetable and mineral—of the integuments, so to
speak, with which nature has clothed the skeleton framework of the globe.
It is difficult to convey by description an idea of the desolation of the regions most exposed to the ravages of torrent and of flood, and the thousands, who, in these days of travel, are whirled by steam near or even
through the theatres of these calamities, have but rare and imperfect opportunities of observing the destructive causes in action. Still more rarely can
they compare the past with the actual condition of the provinces in question, and trace the progress of their conversion from forest-crowned hills,
luxuriant pasture grounds, and abundant cornfields and vineyards well watered by spring and fertilizing rivulets, to bald mountain ridges, rocky declivities, and steep earth banks furrowed by deep ravines with beds now dry,
now filled by torrents of fluid mud and gravel hurrying down to spread
themselves over the plain, and dooming to everlasting barrenness the once
productive fields. In traversing such scenes, it is difficult to resist the im-
pression that nature pronounced the course of perpetual sterility and desolation upon these sublime but fearful wastes, difficult to believe that they
were once, and but for the folly of man might still be, blessed with all the
natural advantages which Providence has bestowed upon the most favored
climes. But the historical evidence is conclusive as to the destructive
changes occasioned by the agency of man upon the flanks of the Alps, the
Apennines, the Pyrenees, and other mountain ranges in Central and
Southern Europe, and the progress of physical deterioration has been so
rapid that, in some localities, a single generation has witnessed the beginning and the end of the melancholy revolution.

It is certain that a desolation, like that which has overwhelmed many
once beautiful and fertile regions of Europe, awaits an important part of the
territory of the United States, and of other comparatively new countries
over which European civilization is now extending its sway, unless prompt
measures are taken to check the action of destructive causes already in operation. It is vain to expect that legislation can do anything effectual to arrest the progress of the evil in those countries, except so far as the state is
still the proprietor of extensive forests. Woodlands which have passed into
private hands will everywhere be managed, in spite of legal restrictions,
upon the same economical principles as other possessions, and every proprietor will, as a general rule, fell his woods, unless he believes that it will be
for his pecuniary interest to preserve them. Few of the new provinces which
the last three centuries have brought under the control of the European
race, would tolerate any interference by the law-making power with what

they regard as the most sacred of civil rights—the right, namely, of every man to do what he will with his own. In the Old World, even in France, whose people, of all European nations, love best to be governed and are least annoyed by bureaucratic supervision, law has been found impotent to prevent the destruction, or wasteful economy, of private forests; and in many of the mountainous departments of that country, man is at this moment so fast laying waste the face of the earth, that the most serious fears are entertained, not only of the depopulation of those districts, but of enormous mischiefs to the provinces contiguous to them. The only legal provisions from which anything is to be hoped, are such as shall make it a matter of private advantage to the landholder to spare the trees upon his grounds, and promote the growth of the young wood. Something may be done by exempting standing forests from taxation, and by imposing taxes on wood felled for fuel or for timber, something by premiums or honorary distinctions for judicious management of the woods. It would be difficult to induce governments, general or local, to make the necessary appropriations for such purposes, but there can be no doubt that it would be sound economy in the end.

In countries where there exist municipalities endowed with an intelligent public spirit, the purchase and control of forests by such corporations would often prove advantageous; and in some of the provinces of Northern Lombardy, experience has shown that such operations may be conducted with great benefit to all the interests connected with the proper management of the woods. In Switzerland, on the other hand, except in some few cases where woods have been preserved as a defence against avalanches, the forests of the communes have been productive of little advantage to the public interests, and have very generally gone to decay. The rights of pasturage, everywhere destructive to trees, combined with toleration of trespasses, have so reduced their value, that there is, too often, nothing left that is worth protecting. In the canton of Ticino, the peasants have very frequently voted to sell the town woods and divide the proceeds among the corporators. The sometimes considerable sums thus received are squandered in wild revelry, and the sacrifice of the forests brings not even a momentary benefit to the proprietors.

It is evidently a matter of the utmost importance that the public, and especially land owners, be roused to a sense of the dangers to which the indiscriminate clearing of the woods may expose not only future generations, but the very soil itself. Fortunately, some of the American States, as well as the governments of many European colonies, still retain the ownership of great tracts of primitive woodland. The State of New York, for example, has, in its northeastern counties, a vast extent of territory in which the lumberman has only here and there established his camp, and where the forest, though interspersed with permanent settlements, robbed of some of its finest pine groves, and often ravaged by devastating fires, still covers far the largest proportion of the surface. Through this territory, the soil is generally poor, and even the new clearings have little of the luxuriance of harvest which distinguishes them elsewhere. The value of the land for agricultural uses is therefore very small, and few purchases are made for any other purpose than to strip the soil of its timber. It has been often proposed that the State should declare the remaining forest the inalienable property of the commonwealth, but I believe the motive of the suggestion has originated rather in poetical than in economical views of the subject. Both these classes of considerations have a real worth. It is desirable that some large and easily accessible region of American soil should remain, as far as possible, in its primitive condition, at once a museum for the instruction of the student, a garden for the recreation of the lover of nature, and an asylum where indigenous tree, and humble plant that loves the shade, and fish and fowl and four-footed beast, may dwell and perpetuate their kind, in the enjoyment of such imperfect protection as the laws of a people jealous of restraint can afford them. The immediate loss to the public treasury from the adoption of this policy would be inconsiderable, for these lands are sold at low rates. The forest alone, economically managed, would, without injury, and even with benefit to its permanence and growth, soon yield a regular income larger than the present value of the fee.

The collateral advantages of the preservation of these forests would be far greater. Nature threw up those mountains and clothed them with lofty woods, that they might serve as a reservoir to supply with perennial waters the thousand rivers and rills that are fed by the rains and snows of the Adirondacks, and as a screen for the fertile plains of the central counties against the chilling blasts of the north wind, which meet no other barrier in their sweep from the Arctic pole. The climate of Northern New York even now presents greater extremes of temperature than that of Southern France. The long continued cold of winter is far more intense, the short heats of summer not less fierce than in Provence, and hence the preservation of every influence that tends to maintain an equilibrium of temperature and humidity is of cardinal importance. The felling of the Adirondack woods would ultimately involve for Northern and Central New York consequences similar to those which have resulted from the laying bare of the southern and western declivities of the French Alps and the spurs, ridges, and detached peaks in front of them.

It is true that the evils to be apprehended from the clearing of the mountains of New York may be less in degree than those which a similar cause has produced in Southern France, where the intensity of its action has been increased by the inclination of the mountain declivities, and by the peculiar geological constitution of the earth. The degradation of the soil is, perhaps, not equally promoted by a combination of the same circumstances, in any of the American Atlantic States, but still they have rapid slopes and loose and friable soils enough to render widespread desolation certain, if the further destruction of the woods is not soon arrested. The effects of clearing are already perceptible in the comparatively unviolated region of which I am speaking. The rivers which rise in it, flow with diminished currents in dry seasons, and with augmented volumes of water after heavy rains. They bring down much larger quantities of sediment, and the increasing obstructions to the navigation of the Hudson, which are extending themselves down the channel in proportion as the fields are encroaching upon the forest, give good grounds for the fear of serious injury to the commerce of the important towns on the upper waters of that river, unless measures are taken to prevent the expansion of "improvements" which have already been carried beyond the demands of a wise economy....

There is one effect of the action of torrents which few travellers on the Continent are heedless enough to pass without notice. I refer to the elevation of the beds of mountain streams in consequence of the deposit of the debris with which they are charged. To prevent the spread of sand and gravel over the fields and the deluging overflow of the raging waters, the streams are confined by walls and embankments, which are gradually built higher and higher as the bed of the torrent is raised, so that, to reach a river, you ascend from the fields beside it; and sometimes the ordinary level of the stream is above the streets and even the roofs of the towns through which it passes.

The traveller who visits the depths of an Alpine ravine, observes the length and width of the gorge and the great height and apparent solidity of the precipitous walls which bound it, and calculates the mass of rock required to fill the vacancy, can hardly believe that the humble brooklet which purls at his feet has been the principal agent in accomplishing this tremendous erosion. Closer observation will often teach him, that the seemingly unbroken rock which overhangs the valley is full of cracks and fissures, and really in such a state of disintegration that every frost must bring down tons of it. If he compute the area of the basin which finds here its only discharge, he will perceive that a sudden thaw of the winter's deposit of snow, or one of those terrible discharges of rain so common in the Alps, must send forth a deluge mighty enough to sweep down the largest masses of gravel and of rock....

I do not mean to assert that all the rocky valleys of the Alps have been produced by the action of torrents resulting from the destruction of the forests. All the greater, and many of the smaller channels, by which that chain is drained, owe their origin to higher causes. They are primitive fissures, ascribable to disruption in upheaval or other geological convulsion, widened and scarped, and often even polished, so to speak, by the action of glaciers during the ice period, and but little changed in form by running water in later eras.

. . . The torrent-worn ravines, of which I have spoken, are of later date, and belong more properly to what may be called the crust of the Alps, consisting of loose rocks, of gravel, and of earth, strewed along the surface of the great declivities of the central ridge, and accumulated thickly between their solid buttresses. But it is on this crust that the mountaineer dwells. Here are his forests, here his pastures, and the ravages of the torrent both destroy his world, and convert it into a source of overwhelming desolation to the plains below.

Transporting Power of Rivers

An instance that fell under my own observation in 1857, will serve to show something of the eroding and transporting power of streams which, in these respects, fall incalculably below the torrents of the Alps. In a flood of the Ottaquechee, a small river which flows through Woodstock, Vermont, a milldam on that stream burst, and the sediment with which the pond was filled, estimated after careful measurement at 13,000 cubic yards, was carried down by the current. Between this dam and the slack water of another, four miles, the bed of the stream, which is composed of pebbles interspersed in a few places with larger stones, is about sixty-five feet wide, though, at low water, the breadth of the current is considerably less. The sand and fine gravel were smoothly and evenly distributed over the bed to a width of fifty-five or sixty feet, and for a distance of about two miles, except at two or three intervening rapids, filling up all the interstices between the stones, covering them to the depth of nine or ten inches, so as to present a regularly formed concave channel, lined with sand, and reducing the depth of water, in some places, from five or six feet to fifteen or eighteen inches. Observing this deposit after the river had subsided and become so clear that the bottom could be seen, I supposed that the next flood would produce an extraordinary erosion of the banks, and some permanent changes in the channel of the stream, in consequence of the elevation of the bed and the filling up of the spaces between the stones through which formerly much water had flowed; but no such result followed. The spring freshet of the next year entirely washed out the sand its predecessor had deposited, carried it to ponds and still-water reaches below, and left the bed of the river almost precisely in its former condition, though, of course, with the slight displacement of the pebbles which every flood produces in the channels of such streams. The pond, though often previously discharged by the breakage of the dam, had then been undisturbed for about twenty-five years, and its contents consisted almost entirely of sand, the rapidity of the current in floods being such that it would let fall little lighter sediment, even above an obstruction like a dam. The quantity I have mentioned evidently bears a very inconsiderable proportion to the total erosion of the stream during that period, because the wash of the banks consists chiefly of fine earth rather than of sand, and after the pond was once filled, or nearly so, even this material could no longer be deposited in it. The fact of the complete removal of the deposit I have described between the two dams in a single freshet, shows that, in spite of considerable obstruction from roughness of bed, large quantities of sand may be taken up and carried off by streams of no great rapidity of inclination; for the whole descent of the bed of the river between the two dams—a distance of four miles—is but sixty feet, or fifteen feet to the mile.

The Po and its Deposits

... We cannot measure the share which human action has had in augmenting the intensity of causes of mountain degradation, but we know that the clearing of the woods has, in some cases, produced within two or three generations, effects as blasting as those generally ascribed to geological convulsions, and has laid waste the face of the earth more hopelessly than if it had been buried by a current of lava or a shower of volcanic sand. New torrents are forming every year in the Alps. Tradition, written records, and analogy concur to establish the belief that the ruin of most of the now desolate valleys in those mountains is to be ascribed to the same cause, and authentic descriptions of the irresistible force of the torrent show that, aided by frost and heat, it is adequate to level Mont Blanc and Monte Rosa themselves, unless new upheavals shall maintain their elevation.

It has been contended that all rivers which take their rise in mountains originated in torrents. These, it is said, have lowered the summits by gradual erosion, and, with the material thus derived, have formed shoals in the sea which once beat against the cliffs; then, by successive deposits, gradually raised them above the surface, and finally expanded them into broad plains traversed by gently flowing streams. If we could go back to earlier geological periods, we should find this theory often verified, and we cannot fail to see that the torrents go on at the present hour, depressing still lower the ridges of the Alps and the Apennines, raising still higher the plains of Lombardy and Provence, extending the coast still farther into the Adriatic and the Mediterranean, reducing the inclination of their own beds and the rapidity of their flow, and thus tending to become river-like in character.

There are cases where torrents cease their ravages of themselves, in consequence of some change in the condition of the basin where they originate, or of the face of the mountain at a higher level, while the plain or the sea below remains in substantially the same state as before. If a torrent rises in a small valley containing no great amount of earth and of disintegrated or loose rock, it may, in the course of a certain period, wash out all the transportable material, and if the valley is then left with solid walls, it will cease to furnish debris to be carried down by floods. If, in this state of things, a new channel be formed at an elevation above the head of the valley, it may divert a part, or even the whole of the rain water and melted snow which would otherwise have flowed into it, and the once furious torrent now sinks to the rank of a humble and harmless brooklet. . . .

But for the intervention of man and domestic animals, these latter beneficent revolutions would occur more frequently, proceed more rapidly. The new scarped mountains, the hillocks of debris, the plains elevated by sand and gravel spread over them, the shores freshly formed by fluviatile deposits, would clothe themselves with shrubs and trees, the intensity of the causes of degradation would be diminished, and nature would thus regain her ancient equilibrium. But these processes, under ordinary circumstances, demand, not years, generations, but centuries; and man, who even now finds scarce breathing room on this vast globe, cannot retire from the Old World to some yet undiscovered continent, and wait for the slow action of such causes to replace, by a new creation, the Eden he has wasted.

Mountain Slides

I have said that the mountainous regions of the Atlantic States of the American Union are exposed to similar ravages, and I may add that there is, in some cases, reason to apprehend from the same cause even more appalling calamities than those which I have yet described. The slide in the Notch of the White Mountains, by which the Willey family lost their lives, is an instance of the sort I refer to, though I am not able to say that in this particular case, the slip of the earth and rock was produced by the denudation of the surface. It may have been occasioned by this cause, or by the

construction of the road through the Notch, the excavations for which, perhaps, cut through the buttresses that supported the sloping strata above.

Not to speak of the fall of earth when the roots which held it together,
and the bed of leaves and mould which sheltered it both from disintegrating frost and from sudden drenching and, dissolution by heavy showers, are
gone, it is easy to see that, in a climate with severe winters, the removal of
the forest, and consequently, of the soil it had contributed to form, might
cause the displacement and descent of great masses of rock. The woods, the
vegetable mould, and the soil beneath, protect the rocks they cover from
the direct action of heat and cold, and from the expansion and contraction
which accompany them. Most rocks, while covered with earth, contain a
considerable quantity of water. A fragment of rock pervaded with moisture
cracks and splits, if thrown into a furnace, and sometimes with a loud deto-
nation; and it is a familiar observation that the fire, in burning over newly
cleared lands, breaks up and sometimes almost pulverizes the stones. This
effect is due partly to the unequal expansion of the stone, partly to the action of heat on the water it contains in its pores. The sun, suddenly let in
upon rock which had been covered with moist earth for centuries, produces
more or less disintegration in the same way, and the stone is also exposed to
chemical influences from which it was sheltered before. But in the climate
of the United States as well as of the Alps, frost is a still more powerful
agent in breaking up mountain masses. The soil that protects the lime and
sand stone, the slate and the granite from the influence of the sun, also prevents the water which filters into their crevices and between their strata
from freezing in the hardest winters, and the moisture descends, in a liquid
form, until it escapes in springs, or passes off by deep subterranean channels. But when the ridges are laid bare, the water of the autumnal rains fills
the minutest pores and veins and fissures and lines of separation of the
rocks, then suddenly freezes, and bursts asunder huge, and apparently solid
blocks of adamantine stone. Where the strata are inclined at a considerable
angle, the freezing of a thin film of water over a large interstratal area might
occasion a slide that should cover miles with its ruins; and similar results
might be produced by the simple hydrostatic pressure of a column of water,
admitted by the removal of the covering of earth to flow into a crevice faster
than it could escape through orifices below.

Earth or rather mountain slides, compared to which the catastrophe that
buried the Willey family in New Hampshire was but a pinch of dust, have
often occurred in the Swiss, Italian, and French Alps. The land slip, which
overwhelmed, and covered to the depth of seventy feet, the town of Plurs in
the valley of the Maira, on the night of the 4th of September, 1618, sparing
not a soul of a population of 2,430 inhabitants, is one of the most memorable

of these catastrophes, and the fall of the Rossberg or Rufiberg, which destroyed the little town of Goldau in Switzerland, and 450 of its people, on the 2d of September, 1806, is almost equally celebrated. In 1771, according to Wessely, the mountain peak Piz, near Alleghe in the province of Belluno, slipped into the bed of the Cordevole, a tributary of the Piave, destroying in its fall three hamlets and sixty lives. The rubbish filled the valley for a distance of nearly two miles, and, by damming up the waters of the Cordevole, formed a lake about three miles along, and a hundred and fifty feet deep, which still subsists, though reduced to half its original length by the wearing down of its outlet.

On the 14th of February, 1855, the hill of Belmonte, a little below the parish of San Stefano, in Tuscany, slid into the valley of the Tiber, which consequently flooded the village to the depth of fifty feet, and was finally drained off by a tunnel. The mass of debris is stated to have been about 3,500 feet long, 1,000 wide, and not less than 600 high.

Such displacements of earth and rocky strata rise to the magnitude of geological convulsions, but they are of so rare occurrence in countries still covered by the primitive forest, so common where the mountains have been stripped of their native covering, and, in many cases, so easily explicable by the drenching of incohesive earth from rain, or the free admission of water between the strata of rocks—both of which a coating of vegetation would have prevented—that we are justified in ascribing them for the most part to the same cause as that to which the destructive effects of mountain torrents are chiefly due—the felling of the woods. . . .

Principal Causes of the Destruction of the Forest

The needs of agriculture are the most familiar cause of the destruction of the forest in new countries; for not only does an increasing population demand additional acres to grow the vegetables which feed it and its domestic animals, but the slovenly husbandry of the border settler soon exhausts the luxuriance of his first fields, and compels him to remove his household gods to fresher soil. With growing numbers, too, come the many arts for which wood is the material. The demands of the near and the distant market for this product excite the cupidity of the hardy forester, and a few years of that wild industry of which Springer's "Forest Life and Forest Trees" so vividly depicts the dangers and the triumphs, suffice to rob the most inaccessible glens of their fairest ornaments. The value of timber increases with its dimensions in almost geometrical proportion, and the tallest, most vigorous, and most symmetrical trees fall the first sacrifice. This is a fortunate circumstance for the remainder of the wood; for the impatient lumberman contents himself with felling a few of the best trees, and then hurries on to take his tithe of still virgin groves.

The unparalleled facilities for internal navigation, afforded by the numerous rivers of the present and former British colonial possessions in North America, have proved very fatal to the forests of that continent. Quebec has become a centre for a lumber trade, which, in the bulk of its material, and, consequently, in the tonnage required for its transportation, rivals the commerce of the greatest European cities. Immense rafts are collected at Quebec from the great Lakes, from the Ottawa, and from all the other tributaries which unite to swell the current of the St. Lawrence and help it to struggle against its mighty tides. Ships, of burden formerly undreamed of, have been built to convey the timber to the markets of Europe, and during the summer months the St. Lawrence is almost as crowded with the vessels as the Thames. Of late, Chicago, in Illinois, has been one of the greatest lumber as well as grain depots of the United States, and it receives and distributes contributions from all the forests in the States washed by Lake Michigan, as well as from some more distant points.

The operations of the lumberman involve other dangers to the woods besides the loss of the trees felled by him. The narrow clearings around his shanties form openings which let in the wind, and thus sometimes occasion the overthrow of thousands of trees, the fall of which dams up small streams, and creates bogs by the spreading of the waters, while the decaying trunks facilitate the multiplication of the insects which breed in dead wood, and are, some of them, injurious to living trees. The escape and spread of camp fires, however, is the most devastating of all the causes of destruction that find their origin in the operations of the lumberman. The proportion of trees fit for industrial uses is small in all primitive woods. Only these fall before the forester's axe, but the fire destroys, indiscriminately, every age and every species of tree. While, then, without much injury to the younger growths, the native forest will bear several "cuttings over" in a generation— for the increasing value of lumber brings into use, every four or five years, a quality of timber which had been before rejected as unmarketable—a fire may render the declivity of a mountain unproductive for a century.

(Between fifty and sixty years ago, a steep mountain with which I am very familiar, composed of metamorphic rock, and at that time covered with a thick coating of soil and a dense primeval forest, was accidentally burnt over. The fire took place in a very dry season, the slope of the mountain was too rapid to retain much water, and the conflagration was of an extraordinarily fierce character, consuming the wood almost entirely, burning the leaves and combustible portion of the mould, and in many places cracking and disintegrating the rock beneath. The rains of the following autumn carried off much of the remaining soil, and the mountain side was nearly bare of wood for two or three years afterward. At length, a new crop of trees sprang up and grew vigorously, and the mountain is now thickly covered again. But the depth of mould and earth is too small to allow the trees to reach maturity. When they attain to the diameter of about six inches, they uniformly die, and this they will no doubt continue to do until the decay of leaves and wood on the surface, and the decomposition of the subjacent rock, shall have formed, perhaps hundreds of years hence, a stratum of soil thick enough to support a full-grown forest.)

American Forest Trees

The remaining forests of the Northern States and of Canada no longer boast the mighty pines which almost rivalled the gigantic Sequoia of California; and the growth of the larger forest trees is so slow, after they have attained to a certain size, that if every pine and oak were spared for two centuries, the largest now standing would not reach the stature of hundreds recorded to have been cut within two or three generations. . . .

Dr. Dwight says that a fallen pine in Connecticut was found to measure two hundred and forty-seven feet in height, and adds: "A few years since, such trees were in great numbers along the northern parts of Connecticut River." In another letter, he speaks of the white pine as "six feet in diameter, and frequently two hundred and fifty feet in height," and states that a pine had been cut in Lancaster, New Hampshire, which measured two hundred and sixty-four feet. Emerson wrote in 1846: "Fifty years ago, several trees growing on rather dry land in Blandford, Massachusetts, measured, after they were felled, . . . two hundred and twenty-three feet." All these trees are surpassed by a pine felled at Hanover, New Hampshire, about a hundred years ago, and described as measuring two hundred and seventy-four feet.

. . . For this change in the growth of forest trees there are two reasons: the one is, that the great commercial value of the pine and the oak have caused the destruction of all the best—that is, the tallest and straightest—specimens of both; the other, that the thinning of the woods by the axe of the lumberman has allowed the access of light and heat and air to trees of humbler worth and lower stature, which have survived their more towering brethren. These, consequently, have been able to expand their crowns and swell their stems to a degree not possible so long as they were overshadowed and stifled by the lordly oak and pine. . . .

Another evil, sometimes of serious magnitude, which attends the operations of the lumberman, is the injury to the banks of rivers from the practice of floating. I do not here allude to rafts, which, being under the control
of those who navigate them, may be so guided as to avoid damage to the
shore, but to masts, logs, and other pieces of timber singly intrusted to the
streams, to be conveyed by their currents to sawmill ponds, or to convenient
places for collecting them into rafts. The lumbermen usually haul the timber to the banks of the rivers in the winter, and when the spring floods swell
the streams and break up the ice, they roll the logs into the water, leaving
them to float down to their destination. If the transporting stream is too
small to furnish a sufficient channel for this rude navigation, it is sometimes dammed up, and the timber collected in the pond thus formed above
the dam. When the pond is full, a sluice is opened, or the dam is blown up
or otherwise suddenly broken, and the whole mass of lumber above it is
hurried down with the rolling flood. Both of these modes of proceeding expose the banks of the rivers employed as channels of flotation to abrasion,
and in some of the American States it has been found necessary to protect,
by special legislation, the lands through which they flow from the serious
injury sometimes received through the practices I have described.

Small Forest Plants, and Vitality of Seeds

Another function of the woods to which I have barely alluded deserves a
fuller notice than can be bestowed upon it in a treatise the scope of which is
purely economical. The forest is the native habitat of a large number of
humbler plants, to the growth and perpetuation of which its shade, its humidity, and its vegetable mould appear to be indispensable necessities. We
cannot positively say that the felling of the woods in a given vegetable province would involve the final extinction of the smaller plants which are
found only within their precincts. Some of these, though not naturally
propagating themselves in the open ground, may perhaps germinate and
grow under artificial stimulation and protection, and finally become hardy
enough to maintain an independent existence in very different circumstances from those which at present seem essential to their life.

. . . When a forest old enough to have witnessed the mysteries of the
Druids is felled, trees of other species spring up in its place; and when they,
in their turn, fall before the axe, sometimes even as soon as they have
spread their protecting shade over the surface, the germs which their pre-
decessors had shed years, perhaps centuries before, sprout up, and in due
time, if not choked by other trees belonging to a later stage in the order of

natural succession, restore again the original wood. In these cases, the seeds of the new crop may often have been brought by the wind, by birds, by quadrupeds, or by other causes; but, in many instances, this explanation is not probable.

When newly cleared ground is burnt over in the United States, the ashes are hardly cold before they are covered with a crop of fire weed, a tall herbaceous plant, very seldom seen growing under other circumstances, and often not to be found for a distance of many miles from the clearing. Its seeds, whether the fruit of an ancient vegetation, or newly sown by winds or birds require either a quickening by a heat which raises to a certain high point the temperature of the stratum where they lie buried, or a special pabulum furnished only by the combustion of the vegetable remains that cover the ground in the woods. Earth brought up from wells or other excavations soon produces a harvest of plants often very unlike those of the local flora. . . .

Although, therefore, the destruction of a wood and the reclaiming of the soil to agricultural uses suppose the death of its smaller dependent flora, these revolutions do not exclude the possibility of its resurrection. In a practical view of the subject, however, we must admit that when the woodman fells a tree he sacrifices the colony of humbler growths which had vegetated under its protection. Some wood plants are known to possess valuable medicinal properties, and experiment may show that the number of these is greater than we now suppose. Few of them, however, have any other economical value than that of furnishing a slender pasturage to cattle allowed to roam in the woods; and even this small advantage is far more than compensated by the mischief done to the young trees by browsing animals. Upon the whole, the importance of this class of vegetables, as physic or as food, is not such as to furnish a very popular argument for the conservation of the forest as a necessary means of their perpetuation. More potent remedial agents may supply their place in the *materia medica,* and an acre of grass land yields more nutriment for cattle than a range of a hundred acres of forest. But he whose sympathies with nature have taught him to feel that there is a fellowship between all God's creatures; to love the brilliant ore better than the dull ingot, iodic silver and crystallized red copper better than the shillings and the pennies forged from them by the coiner's cunning; a venerable oak tree than the brandy cask whose staves are split out from its heart wood; a bed of anemones, hepaticas, or wood violets than the leeks and onions which he may grow on the soil they have enriched and in the air they made fragrant—he who has enjoyed that special training of the heart and intellect which can be acquired only in the unviolated sanctuaries of nature, "where man is distant, but God is near"—will not rashly assert his right to extirpate a tribe of harmless vegetables, barely because their products neither tickle his palate nor fill his pocket; and his regret at the dwindling area of the forest solitude will be augmented by the reflection that the nurselings of the woodland perish with the pines, the oaks, and the beeches that sheltered them.

Although, as I have said, birds do not frequent the deeper recesses of the wood, yet a very large proportion of them build their nests in trees, and find in their foliage and branches a secure retreat from the inclemencies of the seasons and the pursuit of the reptiles and quadrupeds which prey upon them. The borders of the forests are vocal with song; and when the gray morning calls the creeping things of the earth out of their night cells, it summons from the neighboring wood legions of their winged enemies, which swoop down upon the fields to save man's harvests by devouring the destroying worm, and surprising the lagging beetle in his tardy retreat to the dark cover where he lurks through the hours of daylight.

The insects most injurious to rural industry do not multiply in or near the woods. The locust, which ravages the East with its voracious armies, is bred in vast open plains which admit the full heat of the sun to hasten the hatching of the eggs, gather no moisture to destroy them, and harbor no bird to feed upon the larvæ. It is only since the felling of the forests of Asia Minor and Cyrene that the locust has become so fearfully destructive in those countries; and the grasshopper, which now threatens to be almost as great a pest to the agriculture of some North American soils, breeds in seriously injurious numbers only where a wide extent of surface is bare of woods.

Utility of the Forest

In most parts of Europe, the woods are already so nearly extirpated that the mere protection of those which now exist is by no means an adequate remedy for the evils resulting from the want of them; and besides, as I have already said, abundant experience has shown that no legislation can secure the permanence of the forest in private hands. Enlightened individuals in most European states, governments in others, have made very extensive plantations, and France has now set herself energetically at work to restore the woods in the southern provinces, and thereby to prevent the utter depopulation and waste with which that once fertile soil and delicious climate are threatened.

The objects of the restoration of the forest are as multifarious as the motives that have led to its destruction, and as the evils which that destruction has occasioned. It is hoped that the planting of the mountains will diminish the frequency and violence of river inundations, prevent the formation of torrents, mitigate the extremes of atmospheric temperature, humidity, and precipitation, restore dried-up springs, rivulets, and sources of irrigation, shelter the fields from chilling and from parching winds, prevent the spread of miasmatic effluvia, and, finally, furnish an inexhaustible and self-renewing supply of a material indispensable to so many purposes of domestic comfort, to the successful exercise of every art of peace, every destructive energy of war. . . .

Sylviculture

. . . One of the most important rules in the administration of the forest is the absolute exclusion of domestic quadrupeds from every wood which is not destined to be cleared. No growth of young trees is possible where cattle are admitted to pasture at any season of the year, though they are undoubtedly most destructive while trees are in leaf. (Although the economy of the forest has received little attention in the United States, no lover of American nature can have failed to observe a marked difference between a native wood from which cattle are excluded and one where they are permitted to browse. A few seasons suffice for the total extirpation of the "underbrush," including the young trees on which alone the reproduction of the forest depends, and all the branches of those of larger growth which hang within reach of the cattle are stripped of their buds and leaves, and soon wither and fall off.)

It is often necessary to take measures for the protection of young trees against the rabbit, the mole, and other rodent quadrupeds, and of older ones against the damage done by the larvæ of insects hatched upon the surface or in the tissues of the bark, or even in the wood itself. The much greater liability of the artificial than of the natural forest to injury from this cause is perhaps the only point in which the superiority of the former to the latter is not as marked as that of any domesticated vegetable to its wild representative. But the better quality of the wood and the much more rapid growth of the trained and regulated forest are abundant compensations for the loss thus occasioned, and the progress of entomological science will, perhaps, suggest new methods of preventing the ravages of insects. Thus far, however, the collection and destruction of the eggs, by simple but expensive means, has proved the only effectual remedy.

It is common in Europe to permit the removal of the fallen leaves and fragments of bark and branches with which the forest soil is covered, and sometimes the cutting of the lower twigs of evergreens. The leaves and twigs are principally used as litter for cattle, and finally as manure, the bark and wind-fallen branches as fuel. By long usage sometimes by express grant, this privilege has become a vested right of the population in the neighborhood of many public, and even large private forests; but it is generally regarded as a serious evil. To remove the leaves and fallen twigs is to withdraw much of the pabulum upon which the tree was destined to feed. The small branches and leaves are the parts of the tree which yield the largest proportion of ashes on combustion, and of course they supply a great amount of nutriment for the young shoots. . . .

Besides these evils, the removal of the leaves deprives the soil of that spongy character which gives it such immense value as a reservoir of moisture and a regulator of the flow of springs; and, finally, it exposes the surface roots to the drying influence of sun and wind, to accidental mechanical injury from the tread of animals or men, and, in cold climates, to the destructive effects of frost.

The annual lopping and trimming of trees for fuel, so common in Europe, is fatal to the higher uses of the forest, but where small groves are made, or rows of trees planted, for no other purpose than to secure a supply of firewood, or to serve as supports for the vine, it is often very advantageous. The willows, and many other trees, bear polling for a long series of years without apparent diminution of growth of branches, and though certainly a polled, or, to use an old English word, a doddered tree, is in general a melancholy object, yet it must be admitted that the aspect of some species—the American locust, *Robinia pseudacacia*, for instance—when young, is improved by this process.

I have spoken of the needs of agriculture as a principal cause of the destruction of the forest, and of domestic cattle as particularly injurious to the growth of young trees. But these animals affect the forest, indirectly, in a still more important way, because the extent of cleared ground required for agricultural use depends very much on the number and kinds of the cattle bred. We have seen, in a former chapter, that, in the United States, the domestic quadrupeds amount to more than a hundred millions, or three times the number of the human population of the Union. In many of the Western States, the swine subsist more or less on acorns, nuts, and other products of the woods, and the prairies, or natural meadows of the Mississippi valley, yield a large amount of food for beast, as well as for man. With these exceptions, all this vast army of quadrupeds is fed wholly on grass, grain, pulse, and roots grown on soil reclaimed from the forest by European settlers. It is true that the flesh of domestic quadrupeds enters very largely into the aliment of the American people, and greatly reduces the quantity of vegetable nutriment which they would otherwise consume, so that a smaller amount of agricultural product is required for immediate human food, and, of course, a smaller extent of cleared land is needed for the growth of that product, than if no domestic animals existed. But the flesh of the horse, the ass, and the mule is not consumed by man, and the sheep is reared rather for its fleece than for food. Besides this, the ground required to produce the grass and grain consumed in rearing and fattening a grazing quadruped, would yield a far larger amount of nutriment, if devoted to the growing of breadstuffs, than is furnished by his flesh; and, upon the whole, whatever advantages may be reaped from the breeding of domestic cattle, it is plain that the cleared land devoted to their sustenance in the originally wooded part of the United States, after deducting a quantity sufficient to produce an amount of aliment equal to their flesh, still greatly exceeds that culti- vated for vegetables, directly consumed by the people of the same regions; or, to express a nearly equivalent idea in other words, the meadow and the pasture, taken together, much exceed the plough land.

In fertile countries, like the United States, the foreign demand for ani- mal and vegetable aliment, for cotton, and for tobacco, much enlarges the sphere of agricultural operations, and, of course, prompts further encroach- ments upon the forest. The commerce in these articles, therefore, consti- tutes in America a special cause of the destruction of the woods, which does not exist in the numerous states of the Old World that derive the raw material of their mechanical industry from distant lands, and import many articles of vegetable food or luxury which their own climates cannot advan- tageously produce.

The growth of arboreal vegetation is so slow that, though he who buries an acorn may hope to see it shoot up to a miniature resemblance of the ma- jestic tree which shall shade his remote descendants, yet the longest life hardly embraces the seedtime and the harvest of a forest. The planter of a wood must be actuated by higher motives than those of an investment the profits of which consist in direct pecuniary gain to himself or even to his posterity; for if, in rare cases, an artificial forest may, in two or three gener- ations, more than repay its original cost, still, in general, the value of its timber will not return the capital expended and the interest accrued. But when we consider the immense collateral advantages derived from the presence, the terrible evils necessarily resulting from the destruction of the forest, both the preservation of existing woods, and the far more costly ex- tension of them where they have been unduly reduced, are among the most obvious of the duties which this age owes to those that are to come after it. Especially is this obligation incumbent upon Americans. No civilized peo- ple profits so largely from the toils and sacrifices of its immediate predeces- sors as they; no generations have ever sown so liberally, and, in their own persons, reaped so scanty a return, as the pioneers of Anglo-American social life. We can repay our debt to our noble forefathers only by a like magnanimity, by a like self-forgetting care for the moral and material interests of our own posterity.

Instability of American Life

All human institutions, associate arrangements, modes of life, have their characteristic imperfections. The natural, perhaps the necessary defect of ours, is their instability, their want of fixedness, not in form only, but even in spirit. The face of physical nature in the United States shares this incessant fluctuation, and the landscape is as variable as the habits of the population. It is time for some abatement in the restless love of change which characterizes us, and makes us almost a nomade rather than a sedentary people. We have now felled forest enough everywhere, in many districts far too much. Let us restore this one element of material life to its normal proportions, and devise means for maintaining the permanence of its relations to the fields, the meadows, and the pastures, to the rain and the dews of heaven, to the springs and rivulets with which it waters the earth. The establishment of an approximately fixed ratio between the two most broadly characterized distinctions of rural surface—woodland and plough land— would involve a certain persistence of character in all the branches of industry, all the occupations and habits of life, which depend upon or are immediately connected with either, without implying a rigidity that should exclude flexibility of accommodation to the many changes of external circumstance which human wisdom can neither prevent nor foresee, and would thus help us to become, more emphatically, a well-ordered and stable commonwealth, and, not less conspicuously, a people of progress.

Chapter IV: The Waters

Marsh's perspectives on humanity's influence on aquatic ecosystems were narrowed somewhat by the limitations of scientific knowledge in the nineteenth century. Much more was known about terrestrial systems and the immediate land-water interface than of oceans. Therefore, his analyses emphasized modifications of shorelines, rivers, and small bodies of water. Furthermore, the human population in 1864 was only about 1.4 billion people, less than one-fourth of what it is today, and therefore the impact of humans on aquatic ecosystems at that time was much less. Still, he correctly identified a number of examples of changes in aquatic ecosystems, and, as was his philosophy, looked at them in terms of both their positive and negative impacts. He viewed favorably improvements in harbors, diking to reclaim land from the sea, some draining of lakes and swamps to improve public health, some irrigation to support agriculture, and reforestation of mountain slopes to prevent flooding. However, he was aware of such problems as erosion and redeposition of shorelines as a results of human constructions, shorelines sinking from changes in sediment loads in rivers, increased flooding from the construction of reservoirs, and declines in groundwater supplies from deforestation.

Since Marsh's time, both the magnitude of humanity's impacts on aquatic ecosystems and our understanding of them have increased dramatically. The extent of our alteration of the world's waters is such that aquatic ecosystems are now recognized as among the most critically threatened in the world, and the list of environmental problems associated with water is mind-numbing. Since European colonization, more than 50% of the wetlands in the continental United States have been lost to draining, resulting in a frightening decline in water-dwelling plants and animals. Groundwater, vital for both agriculture and drinking water, is being depleted worldwide; for example, the Ogallala aquifer, which supplies water to the American Midwest, is being depleted at a rate 240 times faster than it is replenished. High nutrient loads in rivers, derived from agricultural fertilizers, are creating algal blooms and depleting oxygen levels in oceans and seas. In the Gulf of Mexico, for example, a "dead zone" larger than 7,000 square miles now exists just beyond the mouth of the Mississippi River. The rates of extinction and endangerment of native species in rivers and estuaries are accelerating due to pollution, sedimentation, and the global spread of exotic species in the ballast tanks of oceangoing vessels. Increases in global temperatures and pollution are responsible for the loss of symbiotic

algae in tropical corals, leading to coral decline and death. Fish populations in virtually all the major marine fishery zones have now been overharvested, in many
cases beyond the point of their being able to recover even if a total ban on fish catch
were implemented. Diversion of water from rivers and streams for agricultural and
municipal uses has significantly transformed major bodies of water around the
world; for example, the Colorado River, which at its peak flows at a rate of 635 billion cubic feet per year, now barely even reaches the Golfo de California, and the
Aral Sea in central Asia, once the fourth largest lake in the world, has declined
from its original area of 26,000 square miles by more than half, and from its original volume by more than 80%. Supplies of freshwater are now serious concerns for
more than 5% of people worldwide, a number that is expected to increase rapidly in
the next few decades.

   With respect to water, Marsh was perhaps too optimistic. Evidence suggests
that environmental problems in all aquatic ecosystems will get much worse before
conditions improve, and aquatic conservation remains one of the most challenging
and important issues of our time.

Land artificially won from the Waters

Man, as we have seen, has done much to revolutionize the solid
surface of the globe, and to change the distribution and proportions, if not the essential character, of the organisms which inhabit the
land and even the waters. Besides the influence thus exerted upon the life
which peoples the sea, his action upon the land has involved a certain
amount of indirect encroachment upon the territorial jurisdiction of the
ocean. So far as he has increased the erosion of running waters by the destruction of the forest, he has promoted the deposit of solid matter in the
sea, thus reducing its depth, advancing the coast line, and diminishing the
area covered by the waters. He has gone beyond this, and invaded the realm
of the ocean by constructing within its borders wharves, piers, lighthouses,
break-waters, fortresses, and other facilities for his commercial and military
operations; and in some countries he has permanently rescued from tidal
overflow, and even from the very bed of the deep, tracts of ground extensive enough to constitute valuable additions to his agricultural domain.
The quantity of soil gained from the sea by these different modes of acquisition is indeed, too inconsiderable to form an appreciable element in the

comparison of the general proportion between the two great forms of terrestrial surface, land and water; but the results of such operations, considered in their physical and their moral bearings, are sufficiently important to
entitle them to special notice in every comprehensive view of the relations
between man and nature.

There are cases, as on the western shores of the Baltic, where, in consequence of the secular elevation of the coast, the sea appears to be retiring;
others, where, from the slow sinking of the land, it seems to be advancing.
These movements depend upon geological causes wholly out of our reach,
and man can neither advance nor retard them. There are also cases where
similar apparent effects are produced by local oceanic currents, by river deposit or erosion, by tidal action, or by the influence of the wind upon the
waves and the sands of the sea beach. A regular current may drift suspended earth and seaweed along a coast until they are caught by an eddy
and finally deposited out of the reach of further disturbance, or it may
scoop out the bed of the sea and undermine promontories and headlands; a
powerful river, as the wind changes the direction of its flow at its outlet,
may wash away shores and sandbanks at one point to deposit their material
at another; the tide or waves, stirred to unusual depths by the wind, may
gradually wear down the line of coast, or they may form shoals and coast
dunes by depositing the sand they have rolled up from the bottom of the
ocean. These latter modes of action are slow in producing effects sufficiently important to be noticed in general geography, or even to be visible
in the representations of coast line laid down in ordinary maps; but they
nevertheless form conspicuous features in local topography, and they are attended with consequences of great moment to the material and the moral
interests of men.

The forces which produce these results are all in a considerable degree
subject to control, or rather to direction and resistance, by human power, and
it is in guiding and combating them that man has achieved some of his most
remarkable and honorable conquests over nature. The triumphs in question,
or what we generally call harbor and coast improvements, whether we estimate their value by the money and labor expended upon them, or by their
bearing upon the interests of commerce and the arts of civilization, must
take a very high rank among the great works of man, and they are fast assuming a magnitude greatly exceeding their former relative importance. The
extension of commerce and of the military marine, and especially the introduction of vessels of increased burden and deeper draught of water, have imposed upon engineers tasks of a character which a century ago would have
been pronounced, and, in fact, would have been impracticable; but necessity

has stimulated an ingenuity which has contrived means of executing them, and which gives promise of yet greater performance in time to come.

Men have ceased to admire the power which heaped up the great pyramid to gratify the pride of a despot with a giant sepulchre; for many great harbors, many important lines of internal communication, in the civilized world, now exhibit works which surpass the vastest remains of ancient architectural art in mass and weight of matter, demand the exercise of far greater constructive skill, and involve a much heavier pecuniary expenditure than would now be required for the building of the tomb of Cheops. It is computed that the great pyramid, the solid contents of which when complete were about 3,000,000 cubic yards, could be erected for a million of pounds sterling. The breakwater at Cherbourg, founded in rough water sixty feet deep, at an average distance of more than two miles from the shore, contains double the mass of the pyramid, and many a comparatively unimportant railroad has been constructed at twice the cost which would now build that stupendous monument. Indeed, although man, detached from the solid earth, is almost powerless to struggle against the sea, he is fast becoming invincible by it so long as his foot is planted on the shore, or even on the bottom of the rolling ocean; and though on some battle fields between the waters and the land, he is obliged slowly to yield his ground, yet he retreats still facing the foe, and will finally be able to say to the sea: "Thus far shalt thou come and no farther, and here shall thy proud waves be stayed!" ...

b. Draining of Lakes and Marshes

The substitution of steam engines for the feeble and uncertain action of windmills, in driving pumps, has much facilitated the removal of water from the polders, and the draining of lakes, marshes, and shallow bays, and thus given such an impulse to these enterprises, that not less than one hundred and ten thousand acres were reclaimed from the waters, and added to the agricultural domain of the Netherlands, between 1815 and 1858. The most important of these undertakings was the draining of the Lake of Haarlem, and for this purpose some of the most powerful hydraulic engines ever constructed were designed and executed. The origin of this lake is unknown. It is supposed by some geographers to be a part of an ancient bed of the Rhine, the channel of which, as there is good reason to believe, has undergone great changes since the Roman invasion of the Netherlands; by others it is thought to have once formed an inland marine channel, separated from the sea by a chain of low islands, which the sand washed up by the tides has since connected with the mainland and converted into a continuous line of coast. The best authorities, however, find geological evidence that the surface occupied by the lake was originally a marshy tract containing within its limits little solid ground, but many ponds and inlets, and much floating as well as fixed fen.

In consequence of the cutting of turf for fuel, and the destruction of the few trees and shrubs which held the loose soil together with their roots, the ponds are supposed to have gradually extended themselves, until the action of the wind upon their enlarged surface gave their waves sufficient force to overcome the resistance of the feeble barriers which separated them, and to unite them all into a single lake. Popular tradition, it is true, ascribes the formation of the Lake of Haarlem to a single irruption of the sea, at a remote period, and connects it with one or another of the destructive inundations of which the Netherland chronicles describe so many; but on a map of the year 1531, a chain of four smaller waters occupies nearly the ground afterward covered by the Lake of Haarlem, and they have more probably been united by gradual encroachments resulting from the improvident practices above referred to, though no doubt the consummation may have been hastened by floods, and by the neglect to maintain dikes, or the intentional destruction of them, in the long wars of the sixteenth century. . . .

For this reason, and for the sake of the large addition the bottom of the lake would make to the cultivable soil of the state, it was resolved to drain it, and the preliminary steps for that purpose were commenced in the year 1840. The first operation was to surround the entire lake with a ring canal and dike, in order to cut off the communication with the Ij, and to exclude the water of the streams and morasses which discharge themselves into it from the land side. The dike was composed of different materials, according to the means of supply at different points, such as sand from the coast dunes, earth and turf excavated from the line of the ring canal, and floating turf, fascines being everywhere used to bind and compact the mass together. This operation was completed in 1848, and three steam pumps were then employed for five years in discharging the water. . . .

In a country like the United States, of almost boundless extent of sparsely inhabited territory, such an expenditure for such an object would be poor economy. But Holland has a narrow domain, great pecuniary resources, an excessively crowded population, and a consequent need of enlarged room and opportunity for the exercise of industry. Under such circumstances, and especially with an exposure to dangers so formidable, there is no question of the wisdom of the measure. It has already provided homes and occupation for more than five thousand citizens, and furnished a profitable investment for a capital of not less than £400,000 sterling or

$2,000,000, which has been expended in improvements over and above the purchase money of the soil ; and the greater part of this sum, as well as of the cost of drainage, has been paid as a compensation for labor. . . .

. . . There is good reason to believe that before the establishment of a partially civilized race upon the territory now occupied by Dutch, Frisic, and Low German communities, the grounds not exposed to inundation were overgrown with dense woods, that the lowlands between these forests and the sea coasts were marshes, covered and partially solidified by a thick matting of peat plants and shrubs interspersed with trees, and that even the sand dunes of the shore were protected by a vegetable growth which, in a great measure, prevented the drifting and translocation of them.

The present causes of river and coast erosion existed, indeed, at the period in question; but some of them must have acted with less intensity, there were strong natural safeguards against the influence of marine and fresh-water currents, and the conflicting tendencies had arrived at a condition of approximate equilibrium, which permitted but slow and gradual changes in the face of nature. The reduction of the forests around the sources and along the valleys of the rivers by man gave them a more torrential character. The felling of the trees, and the extirpation of the shrubbery upon the fens by domestic cattle, deprived the surface of cohesion and consistence, and the cutting of peat for fuel opened cavities in it, which, filling at once with water, rapidly extended themselves by abrasion of their borders, and finally enlarged to pools, lakes, and gulfs, like the Lake of Haarlem and the northern part of the Zuiderzee. The cutting of the wood and the depasturing of the grasses upon the sand dunes converted them from solid bulwarks against the ocean to loose accumulations of dust, which every sea breeze drove father landward, burying, perhaps, fertile soil and choking up watercourses on one side, and exposing the coast to erosion by the sea upon the other.

Mountain Lakes

It is a common opinion in America that the river meadows, bottoms, or *intervales,* as they are popularly called, are generally the beds of ancient lakes which have burst their barriers and left running currents in place. It was shown by Dr. Dwight, many years ago, that this is very far from being universally true; but there is no doubt that mountain lakes were of much more frequent occurrence in primitive than in modern geography, and there are many chains of such still existing in regions where man has yet little disturbed the original features of the earth. In the long valleys of the

Adirondack range in Northern New York, and in the mountainous parts of Maine, eight, ten, and even more lakes and lakelets are sometimes found in succession, each emptying into the next lower pool, and so all at last into some considerable river. When the mountain slopes which supply these basins shall be stripped of their woods, the augmented swelling of the lakes will break down their barriers, their waters will run off, and the valleys will present successions of flats with rivers running through them, instead of chains of lakes connected by natural canals....

Geographical and Climatic Effects of Aqueducts, Reservoirs, and Canals

Many processes of internal improvement, such as aqueducts for the supply of great cities, railroad cuts and embankments, and the like, divert water from its natural channels and affect its distribution and ultimate discharge. The collecting of the waters of a considerable district into reservoirs, to be thence carried off by means of aqueducts, as, for example, in the forest of Belgrade, near Constantinople, deprives the grounds originally watered by the springs and rivulets of the necessary moisture, and reduces them to barrenness. Similar effects must have followed from the construction of the numerous aqueducts which supplied ancient Rome with such a profuse abundance of water. On the other hand, the filtration of water through the banks or walls of an aqueduct carried upon a high level across low ground, often injured the adjacent soil, and is prejudicial to the health of the neighboring population; and it has been observed in Switzerland, that fevers have been produced by the stagnation of the water in excavations from which earth has been taken to form embankments for railways.

If we consider only the influence of physical improvements on civilized life, we shall perhaps ascribe to navigable canals a higher importance, or at least a more diversified influence, than to any other works of man designed to control the waters of the earth, and to affect their distribution. They bind distant regions together by social ties, through the agency of the commerce they promote; they facilitate the transportation of military stores and engines, and of other heavy material connected with the discharge of the functions of government; they encourage industry by giving marketable value to raw material and to objects of artificial elaboration which would otherwise be worthless on account of the cost of conveyance; they supply from their surplus waters means of irrigation and of mechanical power; and, in many other ways, they contribute much to advance the prosperity and civilization of nations. Nor are they wholly without geographical importance.

They sometimes drain lands by conveying off water which would otherwise stagnate on the surface, and, on the other hand, like aqueducts, they render the neighboring soil cold and moist by the percolation of water through their embankments; they dam up, check, and divert the course of natural currents, and deliver them at points opposite to, or distant from, their original outlets; they often require extensive reservoirs to feed them, thus retaining through the year accumulations of water—which would otherwise run off, or evaporate in the dry season—and thereby enlarging the evaporable surface of the country; and we have already seen that they interchange the flora and the fauna of provinces widely separated by nature. All these modes of action certainly influence climate and the character of terrestrial surface, though our means of observation are not yet perfected enough to enable us to appreciate and measure their effects.

Surface and Under-draining and their Effects

Superficial draining is a necessity in all lands newly reclaimed from the forest. The face of the ground in the woods is never so regularly inclined as to permit water to flow freely over it. There are, even to the hillsides, many small ridges and depressions, partly belonging to the original distribution of the soil, and partly occasioned by irregularities in the growth and deposit of vegetable matter. These, in the husbandry of nature, serve as dams and reservoirs to collect a larger supply of moisture than the spongy earth can at once imbibe. Besides this, the vegetable mould is, even under the most favorable circumstances, slow in parting with the humidity it has accumulated under the protection of the woods, and the infiltration from neighboring forests contributes to keep the soil of small clearings too wet for the advantageous cultivation of artificial crops. For these reasons, surface draining must have commenced with agriculture itself, and there is probably no cultivated district, one may almost say no single field, which is not provided with artificial arrangements for facilitating the escape of superficial water, and thus carrying off moisture which, in the natural condition of the earth, would have been imbibed by the soil.

The beneficial effects of surface drainage, the necessity of extending the fields as population increased, and the inconveniences resulting from the presence of marshes in otherwise improved regions, must have suggested at a very early period of human industry the expediency of converting bogs and swamps into dry land by drawing off their waters; and it would not be long after the introduction of this practice before further acquisition of agricultural territory would be made by lowering the outlet of small ponds and lakes, and adding the ground they covered to the domain of the husbandman. . . .

This practice has been extensively employed at Paris, not merely for carrying off ordinary surface water, but for the discharge of offensive and deleterious fluids from chemical and manufacturing establishments. A well of this sort received, in the winter of 1832 –'33, twenty thousand gallons per day of the foul water from a starch factory, and the same process was largely used in other factories. The apprehension of injury to common and artesian wells and springs led to an investigation on this subject, in behalf of the municipal authorities, by Girard and Parent Duchatelet, in the latter year. . . . [T]he report came to the conclusion that, in consequence of the absolute immobility of these waters, and the relatively small quantity of noxious fluid to be conveyed to them, there was no danger of the diffusion of this latter, if discharged into them. This result will not surprise those who know that, in another work, Duchatelet maintains analogous opinions as to the effect of the discharge of the city sewers into the Seine upon the waters of that river. The quantity of matter delivered by them he holds to be so nearly infinitesimal, as compared with the volume of water of the Seine, that it cannot possibly affect it to a sensible degree. I would, however, advise determined water drinkers living at Paris to adopt his conclusions, without studying his facts and his arguments; for it is quite possible that he may convert his readers to a faith opposite to his own, and that they will finally agree with the poet who held water an "ignoble beverage."

Climatic and Geographical Effects of Surface Draining

When we remove water from the surface, we diminish the evaporation from it, and, of course, the refrigeration which accompanies all evaporation is diminished in proportion. Hence superficial draining ought to be attended with an elevation of atmospheric temperature, and, in cold countries, it might be expected to lessen the frequency of frosts. Accordingly, it is a fact of experience that, other things being equal, dry soils, and the air in contact with them, are perceptibly warmer during the season of vegetation, when evaporation is most rapid, than moist lands and the atmospheric stratum resting upon them. Instrumental observation on this special point has not yet been undertaken on a very large scale, but still we have thermometric data sufficient to warrant the general conclusion, and the influence of drainage in diminishing the frequency of frost appears to be even better established than a direct increase of atmospheric temperature. The steep and dry uplands of the Green Mountain range in New England often escape frosts when the Indian corn harvest on moister grounds, five hundred or even a thousand feet lower, is destroyed or greatly injured by them. The neighborhood of a marsh is sure to be exposed to late spring and early autumnal frosts, but they cease to be feared after it is drained, and this is particularly observable in very cold climates, as, for example, in Lapland.

In England, under-drains are not generally laid below the reach of daily variations of temperature, or below a point from which moisture might be brought to the surface by capillary attraction and evaporated by the heat of the sun. They, therefore, like surface drains, withdraw from local solar action much moisture which would otherwise be vaporized by it, and, at the same time, by drying the soil above them, they increase its effective hygroscopicity, and it consequently absorbs from the atmosphere a greater quantity of water than it did when, for want of under-drainage, the subsoil was always humid, if not saturated. Under-drains, then, contribute to the dryness as well as to the warmth of the atmosphere, and, as dry ground is more readily heated by the rays of the sun than wet, they tend also to raise the mean, and especially the summer temperature of the soil.

So far as respects the immediate improvement of soil and climate, and the increased abundance of the harvests, the English system of surface and subsoil drainage has fully justified the eulogiums of its advocates; but its extensive adoption appears to have been attended with some altogether unforeseen and undesirable consequences, very analogous to those which I have described as resulting from the clearing of the forests. The under-drains carry off very rapidly the water imbibed by the soil from precipitation, and through infiltration from neighboring springs or other sources of supply. Consequently, in wet seasons, or after heavy rains, a river bordered by artificially drained lands receives in a few hours, from superficial and from subterranean conduits, an accession of water which, in the natural state of the earth, would have reached it only by small installments after percolating through hidden paths for weeks or even months, and would have furnished perennial and comparatively regular contributions, instead of swelling deluges, to its channel. Thus, when human impatience rashly substitutes swiftly acting artificial contrivances for the slow methods by which nature drains the surface and superficial strata of a river basin, the original equilibrium is disturbed, the waters of the heavens are no longer stored up in the earth to be gradually given out again, but are hurried out of man's domain with wasteful haste; and while the inundations of the river are sudden and disastrous, its current, when the drains have run dry, is reduced to a rivulet, it ceases to supply the power to drive the machinery for which it was once amply sufficient, and scarcely even waters the herds that pasture upon its margin. (Babinet condemns even the general draining of marshes. "Draining." says he, "has been much in fashion for some years. It has been a special object to dry and fertilize marshy grounds. My opinion has always been that excessive dryness is thus produced, and that other soils in the neighborhood are sterilized in proportion.")

Irrigation and its Climatic and Geographical Effects

. . . In warm countries . . . the effects I have described as usually resulting from the clearing of the forests would very soon follow. In such climates, the rains are inclined to be periodical; they are also violent, and for these reasons the soil would be parched in summer and liable to wash in winter. In these countries, therefore, the necessity for irrigation must soon have been felt, and its introduction into mountainous regions like Armenia must have been immediately followed by a system of terracing, or at least scarping the hillsides. Pasture and meadow, indeed, may be irrigated even when the surface is both steep and irregular, as may be observed abundantly on the Swiss as well as on the Piedmontese slope of the Alps; but in dry climates, plough land and gardens on hilly grounds require terracing, both for supporting the soil and for administering water by irrigation, and it should be remembered that terracing, of itself, even without special arrangements for controlling the distribution of water, prevents or at least checks the flow of rain water, and gives it time to sink into the ground instead of running off over the surface. . . .

The summers in Egypt, in Syria, and in Asia Minor, and Rumelia, are almost rainless. In such climates, the necessity of irrigation is obvious, and the loss of the ancient means of furnishing it readily explains the diminished fertility of most of the countries in question. The surface of Palestine, for example, is composed, in a great measure, of rounded limestone hills, once, no doubt, covered with forest. These were partially removed before the Jewish conquest. When the soil began to suffer from drought, reservoirs to retain the waters of winter were hewn in the rock near the tops of the hills, and the declivities were terraced. So long as the cisterns were in good order, and the terraces kept up, the fertility of Palestine was unsurpassed, but when misgovernment and foreign and intestine war occasioned the neglect or destruction of these works—traces of which still meet the traveller's eye at every step,—when the reservoirs were broken and the terrace walls had fallen down, there was no longer water for irrigation in summer, the rains of winter soon washed away most of the thin layer of earth upon the rocks, and Palestine was reduced almost to the condition of a desert.

The course of events has been the same in Idumæa. The observing traveller discovers everywhere about Petra, particularly if he enters the city by the route of Wadi Ksheibeh, very extensive traces of ancient cultivation, and upon the neighboring ridges are the ruins of numerous cisterns evidently constructed to furnish a supply of water for irrigation. In primitive ages, the precipitation of winter in these hilly countries was, in great part, retained for a time in the superficial soil, first by the vegetable mould of the forests, and then by the artificial arrangements I have described. The water imbibed by the earth was partly taken up by direct evaporation, partly absorbed by vegetation, and partly carried down by infiltration to subjacent strata which gave it out in springs at lower levels, and thus a fertility of soils and a condition of the atmosphere were maintained sufficient to admit of the dense population that once inhabited those now arid wastes. At present, the rain water runs immediately off from the surface and is carried down to the sea, or is drunk up by the sands of the wadis, and the hillsides which once teemed with plenty are bare of vegetation, and seared by the scorching winds of the desert. . . .

The Nile receives not a single tributary in its course through Egypt; there is not so much as one living spring in the whole land, with the exception of a narrow strip of coast, where the annual precipitation is said to amount to six inches, the fall of rain in the territory of the Pharaohs is not two inches in the year. The subsoil of the whole valley is pervaded with moisture by infiltration from the Nile, and water can everywhere be found at the depth of a few feet. Were irrigation suspended, and Egypt abandoned, as in that case it must be, to the operations of nature, there is no doubt that trees, the roots of which penetrate deeply, would in time establish themselves on the deserted soil, fill the valley with verdure, and perhaps at last temper the climate, and even call down abundant rain from the heavens. But the immediate effect of discontinuing irrigation would be, first, an immense reduction of the evaporation from the valley in the dry season, and then a greatly augmented dryness and heat of the atmosphere. Even the almost constant north wind—the strength of which would be increased in consequence of these changes—would little reduce the temperature of the narrow cleft between the burning mountains which hem in the channel of the Nile, so that a single year would transform the most fertile of soils to the most barren of deserts, and render uninhabitable a territory that irrigation makes capable of sustaining as dense a population as has ever existed in any part of the world. Whether man found the valley of the Nile a forest, or such a waste as I have just described, we do not historically know. In either case, he has not simply converted a wilderness into a garden, but has unquestionably produced extensive climatic change.

The fields of Egypt are more regularly watered than those of any other country bordering on the Mediterranean, except the rice grounds in Italy, and perhaps the *marcite* or winter meadows of Lombardy; but irrigation is more or less employed throughout almost the entire basin of that sea, and is everywhere attended with effects which, if less in degree, are analogous in character to those resulting from it in Egypt. In general, it may be said that the soil is nowhere artificially watered except when it is so dry that little moisture would be evaporated from it, and, consequently, every acre of irrigated ground is so much added to the evaporable surface of the country. When the supply of water is unlimited, it is allowed, after serving its purpose on one field, to run into drains, canals, or rivers. But in most regions where irrigation is regularly employed, it is necessary to economize the water; after passing over or through one parcel of ground, it is conducted to another; no more is withdrawn from the canals at any one point than is absorbed by the soil it irrigates, or evaporated from it, and, consequently, it is not restored to liquid circulation, except by infiltration or precipitation. We are safe, then, in saying that the humidity evaporated from any artificially watered soils is increased by a quantity bearing a large proportion to the whole amount distributed over it; for most even of that which is absorbed by the earth is immediately given out again either by vegetables or by evaporation. . . .

The quantity of water artificially withdrawn from running streams for the purpose of irrigation is such as very sensibly to affect their volume, and it is, therefore, an important element in the geography of rivers. Brooks of no trifling current are often wholly diverted from their natural channels to supply the canals, and their entire mass of water completely absorbed, so that it does not reach the river which it naturally feeds, except in such proportion as it is conveyed to it by infiltration. Irrigation, therefore, diminishes great rivers in warm countries by cutting off their sources of supply as well as by direct abstraction of water from their channels. [T]he system of irrigation in Lombardy deprives the Po of a quantity of water equal to the total delivery of the Seine at ordinary flood, or, in other words, of the equivalent of a tributary navigable for hundreds of miles by vessels of considerable burden. The new canals commenced and projected will greatly increase the loss. The water required for irrigation in Egypt is less than would be supposed from the exceeding rapidity of evaporation in that arid climate; for the soil is thoroughly saturated during the inundation, and infiltration from the Nile continues to supply a considerable amount of humidity in the driest season. . . .

The attentive travellers in Egypt and Nubia cannot fail to notice many localities, generally of small extent, where the soil is rendered infertile by an excess of saline matter in its composition. In many cases, perhaps in all, these barren spots lie rather above the level usually flooded by the inundations of the Nile, and yet they exhibit traces of former cultivation. Recent observations in India, a notice of which I find in an account of a meeting of the Asiatic Society in the Athenæum of December 20, 1862, No. 1834, suggest a possible explanation of this fact. At this meeting, Professor Medlicott read an essay on "the saline efflorescence called 'Reh' and 'Kuller,'" which is gradually invading many of the most fertile districts of Northern and Western India, and changing them into sterile deserts. It consists principally of sulphate of soda (Glauber's salts), with varying proportions of common salt. Mr. Medlicott pronounces "these salts (which, in small quantities are favorable to fertility of soil) to be the gradual result of concentration by evaporation of river and canal waters, which contain them in very minute quantities, and with which the lands are either irrigated or occasionally overflowed." The river inundations in hot countries usually take place but once in a year, and though the banks remain submerged for days or even weeks, the water at that period, being derived principally from rains and snows, must be less highly charged with mineral matter than at lower stages, and besides, it is always in motion. The water of irrigation, on the other hand, is applied for many months in succession, it is drawn from rivers at the seasons when their proportion of salts is greatest, and it either sinks into the superficial soil, carrying with it the saline substances it holds in solution, or is evaporated from the surface, leaving them upon it. Hence irrigation must impart to the soil more salts than natural inundation. The sterilized grounds in Egypt and Nubia lying above the reach of the floods, as I have said, we may suppose them to have been first cultivated in that remote antiquity when the Nile valley received its earliest inhabitants. They must have been artificially irrigated from the beginning; they may have been under cultivation many centuries before the soil at a lower level was invaded by man, and hence it is natural that they should be more strongly impregnated with saline matter than fields which are exposed every year, for some weeks, to the action of running water so nearly pure that it would be more likely to dissolve salts than to deposit them.

Inundations and Torrents: e. Remedies against Inundations

Perhaps no one point has been more prominent in the discussions than the influence of the forest in equalizing and regulating the flow of the water of precipitation. [O]pinion is still somewhat divided on this subject, but the conservative action of the woods in this respect has been generally recognized by the public of France, and the Government of the empire has made this principle the basis of important legislation for the protection of existing forests, and for the formation of new. The clearing of woodland, and the organization and functions of a police for its protection, are regulated by a law bearing date June 18th, 1859, and provision was made for promoting the restoration of private woods by a statute adopted on the 28th of July, 1860. The former of these laws passed the legislative body by a vote of 246 against 4, the latter with but a single negative voice. The influence of the government, in a country where the throne is so potent as in France, would account for a large majority, but when it is considered that both laws, the former especially, interfere very materially with the right of private domain, the almost entire unanimity with which they were adopted is proof of a very general popular conviction, that the protection and extension of the forests is a measure more likely than any other to check the violence, if not to prevent the recurrence, of destructive inundations. The law of July 28th, 1860, appropriated 10,000,000 francs, to be expended, at the rate of 1,000,000 francs per year, in executing or aiding the replanting of woods. It is computed that this appropriation will secure the creation of new forest to the extent of about 250,000 acres, or one eleventh part of the soil where the restoration of the forest is thought feasible and, at the same time, specially important as a security against the evils ascribed in a great measure to its destruction. . . .

Destructive inundations are seldom, if ever, produced by precipitation within the limits of the principal valley, but almost uniformly by sudden thaws or excessive rains on the mountain range where the tributaries take their rise. It is therefore plain that any measures which shall check the flow of surface waters into the channels of the affluents, or which shall retard the delivery of such waters into the principal stream by it tributaries, will diminish in the same proportion the dangers and the evils of inundation by great rivers. The retention of the surface waters upon or in the soil can hardly be accomplished except by the methods already mentioned, replanting of forests, and furrowing or terracing. The current of mountain streams can be checked by various methods, among which the most familiar and obvious is the erection of barriers or dams across their channels, at points convenient for forming reservoirs large enough to retain the superfluous waters of great rains and thaws. Besides the utility of such basins in preventing floods, the construction of them is recommended by very strong considerations, such as the meteorological effect of increased evaporable surface, the furnishing of a constant supply of water for agricultural and mechanical purposes, and, finally, their value as ponds for breeding and rearing fish, and, perhaps, for cultivating aquatic vegetables.

The objections to the general adoption of the system of reservoirs are these: the expense of their construction and maintenance; the reduction of cultivable area by the amount of surface they must cover; the interruption they would occasion to free communication; the probability that they would soon be filled up with sediment, and the obvious fact that when full of earth or even water, they would no longer serve their principal purpose; the great danger to which they would expose the country below them in case of the bursting of their barriers; the evil consequences they would occasion by prolonging the flow of inundations in proportion as they diminished their height; the injurious effects it is supposed they would produce upon the salubrity of the neighboring districts; and, lastly the alleged impossibility of constructing artificial basins sufficient in capacity to prevent, or in any considerable measure to mitigate, the evils they are intended to guard against. . . .

Nor, on the other hand, is this measure to be summarily rejected. Nature has adopted it on a great scale, on both flanks of the Alps, and on a smaller, on those of the Adirondacks and lower chains, and in this as in many other instances, her processes may often be imitated with advantage. The validity of the remaining objections to the system under discussion depends on the topography, geology, and special climate of the regions where it is proposed to establish such reservoirs. Many upland streams present numerous points where none of these objections, except those of expense and of danger from the breaking of dams, could have any application. Reservoirs may be so constructed as to retain the entire precipitation of the heaviest thaws and rains, leaving only the ordinary quantity to flow along the channel; they may be raised to such a height as only partially to obstruct the surface drainage; or they may be provided with sluices by means of which their whole contents can be discharged in the dry season and a summer crop be grown upon the ground they cover at high water. The expediency of employing them and the mode of construction depend on local conditions, and no rules of universal applicability can be laid down on the subject.

It is remarkable that nations which we, in the false pride of our modern civilization, so generally regard as little less than barbarian, should have long preceded Christian Europe in the systematic employment of great artificial basins for the various purposes they are calculated to subserve. The ancient Peruvians built strong walls, of excellent workmanship, across the channels of the mountain sources of important streams, and the Arabs executed immense works of similar description, both in the great Arabian peninsula and in all the provinces of Spain which had the good fortune to fall under their sway. The Spaniards of the fifteenth and sixteenth centuries, who, in many points of true civilization and culture, were far inferior to the races they subdued, wantonly destroyed these noble monuments of social and political wisdom, or suffered them to perish, because they were too ignorant to appreciate their value, or too unskilful as practical engineers to be able to maintain them, and some of their most important territories were soon reduced to sterility and poverty in consequence. . . .

The principal means hitherto relied upon for defense against river inundations has been the construction of dikes along the banks of the streams, parallel to the channel and generally separated from each other by a distance not much greater than the natural width of the bed. If such walls are high enough to confine the water and strong enough to resist its pressure, they secure the lands behind them from all the evils of inundation except those resulting from infiltration; but such ramparts are enormously costly in original construction and maintenance, and, as we have already seen, the filling up of the bed of the river in its lower course, by sand and gravel, involves the necessity of occasionally incurring new expenditures in increasing the height of the banks. They are attended, too, with some collateral disadvantages. They deprive the earth of the fertilizing deposits of the waters, which are powerful natural restoratives of soils exhausted by cultivation; they accelerate the rapidity and transporting power of the current at high water by confining it to a narrower channel, and it consequently conveys to the sea the earthy matter it holds in suspension, and chokes up harbors with a deposit which it would otherwise have spread over a wide surface; they interfere with roads and the convenience of river navigation, and no amount of cost or care can secure them from occasional rupture, in case of which the rush of the waters through the breach is more destructive than the natural flow of the highest inundation.

For these reasons, many experienced engineers are of opinion that the system of longitudinal dikes ought to be abandoned, or, where that cannot be done without involving too great a sacrifice of existing constructions, their elevations should be much reduced, so as to present no obstruction to the lateral spread extraordinary floods, and they should be provided with sluices to admit the water without violence whenever they are likely to be overflowed. Where dikes have not been erected, and where they have been reduced in height, it is proposed to construct, at convenient intervals, transverse embankments of moderate height running from the banks of the river across the plains to the hills which bound them. These measures, it is argued, will diminish the violence of inundations by permitting the waters to extend themselves over a greater surface, and thus retarding the flow of the river currents, and will, at the same time, secure the deposit of fertilizing slime upon all the soil covered by the flood. . . .

The deposit of slime by rivers upon the flats along their banks not only contributes greatly to the fertility of the soil thus flowed, but it subserves a still more important purpose in the general economy of nature. All running streams begin with excavating channels for themselves, or deepening the natural depressions in which they flow; but in proportion as their outlets are raised by the solid material transported by their currents, their velocity is diminished, they deposit gravel and sand at constantly higher and higher points, and so at last elevate, in the middle and lower part of their course, the beds they had previously scooped out. The raising of the channels is compensated in part by the simultaneous elevation of their banks and the flats adjoining them, from the deposit of the finer particles of earth and vegetable mould brought down from the mountains, without which elevation the low grounds bordering all rivers would be, as in many cases they in fact are, mere morasses.

All arrangements which tend to obstruct this process of raising the flats adjacent to the channel, whether consisting in dikes which confine the waters, and, at the same time, augment the velocity of the current, or in other means of producing the last-mentioned effect, interfere with the restorative economy of nature, and at last occasion the formation of marshes where, if left to herself, she would have accumulated inexhaustible stores of the richest soil, and spread them out in plains above the reach of ordinary floods.

Subterranean Waters

I have frequently alluded to a branch of geography, the importance of which is but recently adequately recognized—the subterranean waters of the earth considered as stationary reservoirs, as flowing currents, and as filtrating fluids. The earth drinks in moisture by direct absorption from the atmosphere, by the deposition of dew, by rain and snow, by percolation from rivers and other superficial bodies of water, and sometimes by currents flowing into caves or smaller visible apertures. Some of this humidity is exhaled again by the soil, some is taken up by organic growths and by inorganic compounds, some poured out upon the surface by springs and either immediately evaporated or carried down to larger streams and to the sea, some flows by subterranean courses into the bed of fresh-water rivers or of the ocean, and some remains, though even here not in forever motionless repose, to fill deep cavities and underground channels. In every case the aqueous vapors of the air are the ultimate source of supply, and all these hidden stores are again returned to the atmosphere by evaporation.

The proportion of the water of precipitation taken up by direct evaporation from the surface of the ground seems to have been generally exaggerated, sufficient allowance not being made for moisture carried downward, or in a lateral direction, by infiltration or by crevices in the superior rocky or earthy strata.

According to Wittwer, Mariotte found that but one sixth of the precipitation in the basin of the Seine was delivered into the sea by that river, "so that five sixths remained for evaporation and consumption by the organic world.". . .

Babinet quotes a French proverb, "Summer rain wets nothing," and explains it as meaning that the water of such rains is "almost totally taken up by evaporating." "The rains of summer," he adds, "however abundant they may be, do not penetrate the soil to a greater depth than 15 or 20 cetimètres. In summer the evaporating power of the heat is five or six times as great as in winter, and this power is exerted by an atmosphere capable of containing five times as much vapor as in winter." "A stratum of snow which prevents evaporation [from the soil] causes almost all the water that composes it to filter down into the earth, and form a reserve for springs, wells, and rivers which could not be supplied by any amount of summer rain." "This latter— useful, indeed like dew, to vegetation—does not penetrate the soil and accumulate a store to feed springs and to be brought up by them to the open air." This conclusion, however applicable it may be to the climate and soil of France, is too broadly stated to be accepted as a general truth, and in countries where the precipitation is small in the winter months, familiar observation shows that the quantity of water yielded by deep wells and natural springs depends not less on the rains of summer than on those of the rest of the year, and, consequently, that much of the precipitation of that season must finds its way to strata too deep to lose water by evaporation.

The supply of subterranean reservoirs and currents, as well as of springs, is undoubtedly derived chiefly from infiltration, and hence it must be affected by all changes of the natural surface that accelerate or retard the drainage of the soil, or that either promote or obstruct evaporation from it. It has sufficiently appeared from what has gone before, that the spontaneous drainage of cleared ground is more rapid than that of the forest, and consequently, that the felling of the woods, as well as the draining of swamps, deprives the subterranean waters of accessions which would otherwise be conveyed to them by infiltration. The same effect is produced by artificial contrivances for drying the soil either by open ditches or by underground pipes or channels, and in proportion as the sphere of these operations is extended, the effect of them cannot fail to make itself more and more sensibly felt in the diminished supply of water furnished by wells and running springs.

It is undoubtedly true that loose soils, stripped of vegetation and broken up by the plough or other processes of cultivation, may, until again carpeted by grasses or other plants, absorb more rain and snow water than when they were covered by a natural growth; but it is also true that the evaporation from such soils is augmented in a still greater proportion. Rain scarcely penetrates beneath the sod of grass ground, but runs off over the surface; and after the heaviest showers a ploughed field will often be dried by evaporation before the water can be carried off by infiltration, while the soil of a neighboring grove will remain half saturated for weeks together. Sandy soils frequently rest on a tenacious subsoil, at a moderate depth, as is usually seen in the pine plains of the United States, where pools of rain water collect in slight depression on the surface of earth, the upper stratum of which is as porous as a sponge. In the open grounds such pools are very soon dried up by the sun and wind; in the woods they remain unevaporated long enough for the water to diffuse itself laterally until it finds, in the subsoil, crevices through which it may escape, or slopes which it may follow to their outcrop or descend along them to lower strata.

The readiness with which water not obstructed by impermeable strata diffuses itself through the earth in all directions—and, consequently, the importance of keeping up the supply of subterranean reservoirs—find a familiar illustration in the effect of paving the ground about the stems of vines and trees. The surface earth around the trunk of a tree may be made perfectly impervious to water, by flag stones and cement, for a distance greater than the spread of the roots; and yet the tree will not suffer for want of moisture, except in droughts severe enough sensibly to affect the supply in deep wells and springs. Both forest and fruit trees grow well in cities where the streets and courts are closely paved, and where even the lateral access of water to the roots is more or less obstructed by deep cellars and foundation walls. The deep-lying veins and sheets of water, supplied by infiltration from above, send up moisture by capillary attraction, and the pavement prevents the soil beneath it from losing its humidity by evaporation. Hence, city-grown trees find moisture enough for their roots, and though plagued with smoke and dust, often retain their freshness while those planted in the open fields, where sun and wind dry up the soil faster than the subterranean fountains can water it, are withering from drought. Without the help of artificial conduit or of water carrier, the Thames and the Seine refresh the ornamental trees that shade the thoroughfares of London and of Paris, and beneath the hot and reeking mould of Egypt, the Nile sends currents to the extremest border of its valley.

Chapter V: The Sands

Marsh considered sand dunes as the second great terrestrial ecosystem that humans have modified for both good and ill. He felt that dunes, in general, were "destructive to human industry" except where they protected other lands from the encroachment of the sea. Yet he placed much of the responsibility for the damage that drifting dunes cause on the actions of humans; when dunes are excavated or deforested, or even if the surrounding areas are deforested, the natural conditions that promote the stability of dunes are destroyed. Marsh felt that actions taken to restore the stability of dunes or to restore the agricultural productivity of land destroyed by drifting dunes were admirable.

Our views today about dunes are modified only slightly from those of Marsh. Dunes are still viewed as important physical barriers to wind and wave, whose stability depends on vegetation. We also now recognize that this vegetation is in some cases unique, and that dunes support distinct plant communities made of species adapted to both the soil and moisture conditions. Furthermore, since dunes form and move in response to modified conditions, we view them as important indicators of disturbance and change. Abuse by off-road vehicles and global climate change both are manifested by changes in the distribution and extent of dune ecosystems. For example, since Marsh's time, the southern edge of the Sahara Desert in Africa has advanced at least 200 miles southward in response to forest clearing, overgrazing, and long-term cycles of drought.

Sands of Egypt

In these facts we find the true explanation of the sand drifts, which have half buried the Sphinx and so many other ancient monuments in that part of Egypt. These drifts, as I have said, are not primarily from the desert, but from the sea; and, as might be supposed from the distance they have travelled, they have been long in gathering. While Egypt was a great and flourishing kingdom, measures were taken to protect its territory against the encroachment of sand, whether from the desert or from the sea; but the foreign conquerors, who destroyed so many of its religious monuments, did not spare its public works, and the process of physical degradation undoubtedly began as early as the Persian invasion. The urgent necessity, which has compelled all the successive tyrannies of Egypt to keep up some of the canals and other arrangements for irrigation, was not felt with respect to the advancement of the sands; for their progress was so slow as hardly to be perceptible in the course of a single reign, and long experience has shown that, from the natural effect of the inundations, the cultivable soil of the valley is, on the whole, trenching upon the domain of the desert, not retreating before it.

The oases of the Libyan, as well as of many Asiatic deserts, have no such safeguards. The sands are fast encroaching upon them, and threaten soon to engulf them, unless man shall resort to artesian wells and plantations, or to some other efficient means of checking the advance of this formidable enemy, in time to save these islands of the waste from final destruction.

Accumulations of sand are, in certain cases, beneficial as a protection against the ravages of the sea; but, in general, the vicinity, and especially the shifting of bodies of this material, are destructive to human industry, and hence, in civilized countries, measures are taken to prevent its spread. This, however, can be done only where the population is large and enlightened, and the value of the soil, or of the artificial erections and improvements upon it, is considerable. Hence in the deserts of Africa and of Asia, and the inhabited lands which border on them, no pains are usually taken to check the drifts, and when once the fields, the houses, the springs, or the canals of irrigation are covered or choked, the district is abandoned without a struggle, and surrendered to perpetual desolation.

Age, Character, and Permanence of Dunes

. . . It is a question of much interest, in what degree the naked condition of most dunes is to be ascribed to the improvidence and indiscretion of man. There are, in Western France, extensive ranges of dunes covered with ancient and dense forests, while the recently formed sand hills between them and the sea are bare of vegetation, and are rapidly advancing upon the wooded dunes, which they threaten to bury beneath their drifts. Between the old dunes and the new, there is no discoverable difference in material or in structure; but the modern sand hills are naked and shifting, the ancient, clothed with vegetation and fixed. It has been conjectured that artificial methods of confinement and plantation were employed by the primitive inhabitants of Gaul; and Laval, basing his calculations on the rate of annual movement of the shifting dunes, assigns the fifth century of the Christian era as the period when these processes were abandoned.

. . . In other countries, dunes have spontaneously clothed themselves with forests, and the rapidity with which their surface is covered by various species of sand plants, and finally by trees, where man and cattle and burrowing animals are excluded from them, renders it highly probable that they would, as a general rule, protect themselves, if left to the undisturbed action of natural causes. The sand hills of the Frische Nehrung, on the coast of Prussia, were formerly wooded down to the water's edge, and it was only in the last century that, in consequence of the destruction of their forests, they became moving sands. There is every reason to believe that the dunes of the Netherlands were clothed with trees until after the Roman invasion. The old geographers, in describing these countries, speak of vast forests extending to the very brink of the sea; but drifting coast dunes are first mentioned by the chroniclers of the Middle Ages, and so far as we know they have assumed a destructive character in consequence of the improvidence of man. The history of the dunes of Michigan, so far as I have been able to learn from my own observation, or that of others, is the same. Thirty years ago, when that region was scarcely inhabited, they were generally covered with a thick growth of trees, chiefly pines, and underwood, and there was little appearance of undermining and wash on the lake side, or shifting of the sands, except where the trees had been cut or turned up by the roots. (The sands of Cape Cod were partially, if not completely, covered with vegetation by nature. Dr. Dwight, describing the dunes as they were in 1800, says: "Some of them are covered with beach grass; some fringed with whortleberry bushes; and some tufted with a small and singular growth of oaks. * * * The parts of this barrier, which are covered with whortleberry bushes and with oaks, have been either not at all, or very little blown. The oaks, particularly, appear to be the continuation of the forests originally formed on this spot. * * * They wore all the marks of extreme age; were, in some instances, already decayed, and in others decaying; were hoary with moss, and were deformed by branches broken and wasted, not by violence, but by time.")

Nature, as she builds up dunes for the protection of the sea shore, provides, with similar conservatism, for the preservation of the dunes themselves; so that, without the interference of man, these hillocks would be, not perhaps absolutely perpetual, but very lasting in duration, and very slowly altered in form or position. When once covered with the trees, shrubs, and herbaceous growth adapted to such localities, dunes undergo no apparent change, except the slow occasional undermining of the outer tier, and accidental destruction by the exposure of the interior, from the burrowing of animals, or the upturning of trees with their roots, and all these causes of displacement are very much less destructive when a vegetable covering exists in the immediate neighborhood of the breach.

Before the occupation of the coasts by civilized and therefore destructive man, dunes, at all points where they have been observed, seem to have been protected in their rear by forests, which served to break the force of the winds in both directions, and to have spontaneously clothed themselves with a dense growth of the various plants, grasses, shrubs, and trees, which nature has assigned to such soils. (Bergsöe states that the dunes on the west coast of Jutland were stationary before the destruction of the forests to the east of them. The felling of the tall trees removed the resistance to the lower currents of the westerly winds, and the sands have since buried a great extent of fertile soil.) It is observed in Europe that dunes, though now without the shelter of a forest country behind them, begin to protect themselves as soon as human trespassers are excluded, and grazing animals denied access to them. Herbaceous and arborescent plants spring up almost at once, first in the depressions, and then upon the surface of the sand hills. Every seed that sprouts, binds together a certain amount of sand by its roots, shades a little ground with its leaves, and furnishes food and shelter for still younger or smaller growths. A succession of a very few favorable seasons suffices to bind the whole surface together with a vegetable network, and the power of resistance possessed by the dunes themselves, and the protection they afford to the fields behind them, are just in proportion to the abundance and density of the plants they support.

The growth of the vegetable covering can, of course, be much accelerated by judicious planting and watchful care, and this species of improvement is now carried on upon a vast scale, wherever, the value of land is considerable and the population dense. In the main, the dunes on the coast of the German Sea, notwithstanding the great quantity of often fertile land they cover, and the evils which result from their movement, are, upon the whole, a protective and beneficial agent, and their maintenance is an object of solicitude with the governments and people of the shores they protect.

Drifting of Dune Sands

Besides their importance as a barrier against the inroads of the ocean, dunes are useful by sheltering the cultivated ground behind them from the violence of the sea wind, from salt spray, and from the drifts of beach sand which would otherwise overwhelm them. But the dunes themselves, unless their surface sands are kept moist, and confined by the growth of plants, or at least by a crust of vegetable earth, are constantly rolling inward; and thus, while, on one side, they lay bare the traces of ancient human habitations or other evidences of the social life of primitive man, they are, on the other, burying fields, houses, churches, and converting populous districts into barren and deserted wastes.

Especially destructive are they when, by any accident, a cavity is opened into them to a considerable depth, thereby giving the wind access to the interior, where the sand is thus first dried, and then scooped out and scattered far over the neighboring soil. The dune is now a magazine of sand, no longer a rampart against it, and mischief from this source seems more difficult to resist than from almost any other drift, because the supply of material at the command of the wind, is more abundant and more concentrated than in its original thin and widespread deposits on the beach. The burrowing of conies in the dunes is, in this way, not unfrequently a cause of their destruction and of great injury to the fields behind them. Drifts, and even inland sand hills, sometimes result from breaking the surface of more level sand deposits, far within the range of the coast dunes. Thus we learn from Staring, that one of the highest inland dunes in Friesland owes its origin to the opening of the drift sand by the uprooting of a large oak.

Great as are the ravages produced by the encroachment of the sea upon the western shores of continental Europe, they have been in some degree compensated by spontaneous marine deposits at other points of the coast, and we have seen in a former chapter that the industry of man has reclaimed a large territory from the bosom of the ocean. These latter triumphs are not of recent origin, and the incipient victories which paved the way for them date back perhaps as far as ten centuries. In the mean time, the dunes had been left to the operation of the laws of nature, or rather freed, by human imprudence, from the fetters with which nature had bound them, and it is scarcely three generations since man first attempted to check their destructive movements. As they advanced, he unresistingly yielded and retreated before them, and they have buried under their sandy billows many hundreds of square miles of luxuriant cornfields and vineyards and forests.

Protection of Dunes

The dunes of Holland are sometimes protected from the dashing of the waves by a *revêtement* of stone, or by piles; and the lateral high-water currents, which wash away their base, are occasionally checked by transverse walls running from the foot of the dunes to low-water mark; but the great expense of such constructions has prevented their adoption on a large scale. The principal means relied on for the protection of the sand hills are the planting of their surfaces and the exclusion of burrowing and grazing animals. There are grasses, creeping plants, shrubs of spontaneous growth, which flourish in loose sand, and, if protected, spread over considerable tracts, and finally convert their face into a soil capable of cultivation, or, at least, of producing forest trees. Krause enumerates one hundred and seventy-one plants as native to the coast sands of Prussia, and the observations of Andresen in Jutland carry the number of these vegetables up to two hundred and thirty-four.

Some of these plants, especially the *Arundo arenaria* or *arenosa* or *Psamma* or *Psammophila arenaria* . . . are exclusively confined to sandy soils, and thrive well only in a saline atmosphere. The arundo grows to the height of about twenty-four inches, but sends its strong roots with their many rootlets to a distance of forty or fifty feet. It has the peculiar property of flourishing best in the loosest soil, and a sand shower seems to refresh it as the rain revives the thirsty plants of the common earth. Its roots bind together the dunes, and its leaves protect their surface. When the sand ceases to drift, the arundo dies, its decaying roots fertilizing the sand, and the decomposition of its leaves forming a layer of vegetable earth over it. Then follows a succession of other plants which gradually fit the sand hills, by growth and decay, for forest planting, for pasturage, and sometimes for ordinary agricultural use.

But the protection and gradual transformation of the dunes is not the only service rendered by this valuable plant. Its leaves are nutritious food for sheep and cattle, its seeds for poultry; cordage and netting twine are manufactured from its fibres, it makes a good material for thatching, and its dried roots furnish excellent fuel. These useful qualities, unfortunately, are too often prejudicial to its growth. The peasants feed it down with their cattle, cut it for rope making, or dig it up for fuel, and it has been found necessary to resort to severe legislation to prevent them from bringing ruin upon themselves by thus improvidently sacrificing their most effectual safeguard against the drifting of the sands.

In 1539, a decree of Christian III, king of Denmark, imposed a fine upon persons convicted of destroying certain species of sand plants upon the west coast of Jutland. This ordinance was renewed and made more comprehensive in 1558, and in 1569 the inhabitants of several districts were required, by royal rescript, to do their best to check the sand drifts, though the specific measures to be adopted for the purpose are not indicated. Various laws against stripping the dunes of their vegetation were enacted in the following century, but no active measures were taken for the subjugation of the sand drifts until 1779, when a preliminary system of operation for that purpose was adopted. This consisted in little more than the planting of the *Arundo arenaria* and other sand plants, and the exclusion of animals destructive to these vegetables. Ten years later, plantations of forest trees, which have since proved so valuable a means of fixing the dunes and rendering them productive, were commenced, and have been continued ever since. During this latter period, Brémontier, without any knowledge of what was doing in Denmark, experimented upon the cultivation of forest trees on the dunes of Gascony, and perfected a system, which, with some improvements in matters of detail, is still largely pursued on those shores. The example of Denmark was soon followed in the neighboring kingdom of Prussia, and in the Netherlands. . . .

(Measures were taken for the protection of the dunes of Cape Cod, in Massachusetts, during the colonial period, though I believe they are now substantially abandoned. A hundred years ago, before the valley of the Mississippi, or even the rich plains of Central and Western New York, were opened to the white settler, the value of land was relatively much greater in New England than it is at present, and consequently some rural improvements were then worth making, which would not now yield sufficient returns to tempt the investment of capital. The money and the time required to subdue and render productive twenty acres of sea and on Cape Cod, would buy a "section" and rear a family in Illinois. The son of the Pilgrims, therefore, abandons the sand hills, and seeks a better fortune on the fertile prairies of the West. . . .

Dr. Dwight, who visited Cape Cod in the year 1800, after describing the "beach grass, a vegetable bearing a general resemblance to sedge, but of a light bluish-green, and of a coarse appearance," which "flourishes with a strong and rapid vegetation on the sands," observes that he received "from a Mr. Collins, formerly of Truro, the following information:" "When he lived at Truro, the inhabitants were, under the authority of law, regularly warned in the month of April, yearly, to plant beach grass, as, in other towns of New England, they are warned to repair highways. It was required by the laws of the State, and under the proper penalties for disobedience; being as regular a public tax as any other. The people, therefore, generally attended and performed the labor. The grass was dug in bunches, as it naturally grows; and each bunch divided into a number of smaller ones. These were set out in the sand at distances of three feet. After one row was set, others were placed behind it in such a manner as to shut up the interstices; or, as a carpenter would say, so as to break the joints. * * * When it is once set, it grows and spreads with rapidity. * * * The seeds are so heavy that they bend down the heads of the grass; and when ripe, drop directly down by its side, where they immediately vegetate. Thus in a short time the ground is covered.

"Where this covering is found, none of the sand is blown. On the contrary, it is accumulated and raised continually as snow gathers and rises among bushes, or branches of trees cut and spread upon the earth. Nor does the grass merely defend the surface on which it is planted; but rises, as that rises by new accumulations; and always overtops the sand, however high that may be raised by the wind." . . .)

Under the administration of Reventlov, a little before the close of the last century, the Danish Government organized a regular system of improvement in the economy of the dunes. They were planted with the arundo and other vegetables of similar habits, protected against trespassers, and at last partly covered with forest trees. By these means much waste soil has been converted into arable ground, a large growth of valuable timber obtained, and the further spread of the drifts, which threatened to lay waste the whole peninsula of Jutland, to a considerable extent arrested. . . .

Inland Sand Plains

The inland sand plains of Europe are either derived from the drifting of dunes or other beach sands, or consist of diluvial deposits. As we have seen, when once the interior of a dune is laid open to the wind, its contents are soon scattered far and wide over the adjacent country, and the beach sands, no longer checked by the rampart which nature had constrained them to build against their own encroachments, are also carried to considerable distances from the coast. Few regions have suffered so much from this cause in proportion to their extent, as the peninsula of Jutland. So long as the woods, with which nature had planted the Danish dunes, were spared, they seem to have been stationary, and we have no historical evidence, of an earlier date than the sixteenth century, that they had become in any way injurious. From that period, there are frequent notices of the invasions of cultivated grounds by the sands; and excavations are constantly bringing to light proof of human habitation and of agricultural industry, in former ages, on soils now buried beneath deep drifts from the dunes and beaches of the sea coast.

Extensive tracts of valuable plain land in the Netherlands and in France have been covered in the same way with a layer of sand deep enough to render them infertile, and they can be restored to cultivation only by processes analogous to those employed for fixing and improving the dunes. Diluvial sand plains, also, have been reclaimed by these methods in the Duchy of Austria, between Vienna and the Semmering ridge, in Jutland, and in the great champaign country of Northern Germany, especially the Mark Brandenburg, where artificial forests can be propagated with great ease, and where, consequently, this branch of industry has been pursued on a great scale, and with highly beneficial results, both as respects the supply of forest products and the preparation of the soil for agricultural use.

As a general rule, inland sands are looser, dryer, and more inclined to drift, than those of the sea coast, where the moist and saline atmosphere of the ocean keeps them always more or less humid and cohesive. No shore dunes are so movable as the medanos of Peru . . . or as the sand hills of Poland, both of which seem better entitled to the appellation of sand waves than those of the Sahara or of the Arabian desert. The sands of the valley of the Lower Euphrates—themselves probably of submarine origin, and not derived from dunes—are advancing to the northwest with a rapidity which seems fabulous when compared with the slow movement of the sand hills of Gascony and the Low German coasts. Loftus, speaking of Niliyya, an old Arab town a few miles east of the ruins of Babylon, says that, "in 1848, the sand began to accumulate around it, and in six years, the desert, within a radius of six miles, was covered with little, undulating domes, while the ruins of the city were so buried that it is now impossible to trace their original form or extent." Loftus considers this sand flood as the "vanguard of those vast drifts which, advancing from the southeast, threaten eventually to overwhelm Babylon and Baghdad.". . .

Government Works

There is an important political difference between these latter works *[planting on dunes]* and the diking system of the Netherlandish and German coasts. The dikes originally were, and in modern times very generally have been, private enterprises, undertaken with no other aim than to add a certain quantity of cultivable soil to the former possessions of their proprietor, or sometimes of the state. In short, with few exceptions, they have been merely a pecuniary investment, a mode of acquiring land not economically different from purchase. The planting of the dunes, on the contrary, has always been a public work, executed, not with the expectation of reaping a regular direct percentage of income from the expenditure, but dictated by higher views of state economy—by the same governmental principles, in fact, which animate all commonwealths in repelling invasion by hostile armies, or in repairing the damages that invading forces may have inflicted on the general interests of the people. The restoration of the forests in the southern part of France, as now conducted by the Government of that empire, is a measure of the same elevated character as the fixing of the dunes. In former ages, forests were formed or protected simply for the sake of the shelter they afforded to game, or for the timber they yielded; but the recent legislation of France, and of some other Continental countries, on this subject, looks to more distant as well as nobler ends, and these are among the public acts which most strongly encourage the hope that the rulers of Christendom are coming better to understand the true duties and interests of civilized government.

Chapter VI: Projected or Possible Geographical Changes by Man

In this concluding chapter, Marsh developed two separate themes to draw his exploration of the geographical influences of humanity to a close. The first theme was the impacts, positive and negative, of large-scale projects that moved earth or controlled water. Although he felt that such projects as maritime canals and changes in soil levels could be justified in terms of both commercial value and improvement of human lives, he took a precautionary approach by noting that there were many environmental uncertainties in the development of such projects that should be considered before proceeding. Much of his attention here was given to the Suez Canal because it was undergoing construction while Marsh was writing this book. Completed in 1869, it achieved the commercial goals that were set for it with little or no negative environmental impact. Marsh commented on the possibility of a canal across Central America. The Panama Canal was built from 1904 to 1913; the negative impacts that Marsh warned about should the canal be built at sea level were avoided by constructing it as a series of locks that use only freshwater, thereby preventing the exchange of marine fauna between the Atlantic and Pacific Oceans.

The general success of some of the large-scale projects completed during the late nineteenth and early twentieth centuries should not, however, lead to complacency. Others have been less successful, such as the Glen Canyon Dam on the Colorado River (completed in 1963; flooding, extinction of native species, loss of riverine ecosystems) and Aswan High Dam on the Nile River (completed in 1970; flooding, loss of downstream soil fertility). The lesson should be that large-scale projects can succeed in the long run only if attention is paid to discovering and addressing environmental consequences before construction begins. Unfortunately, this is rarely done. The proposed Three Gorges Dam on the Chang Jiang River in China is a case in point. Currently planned for completion in 2009, this dam will be 1.2 miles long and will flood an area upstream of the dam for 370 miles. Although the project will displace over a million people and is.predicted to destroy hundreds of miles of wildlife habitat, deplete nutrient flow to downstream agricultural land, and increase the damage from earthquakes and landslides, it is still moving forward.

The final theme of this chapter, and of this book, is one of humility. In the book's concluding paragraph, Marsh offers one of the most eloquent statements ever written about the relationship between humanity and nature. He asks us to realize or remember that nature knows no trifles; her laws are inflexible, and it is our own limitations that make it difficult to understand the ultimate consequences of our actions. Our ignorance is not an excuse for ignoring the possibility that our actions might have devastating consequences. Our ignorance can be cured only through a study of nature, and through this study we may come to answer the question of whether humanity is "of nature or above her." He thus set the stage for more than a hundred years of discussion among philosophers, writers, religious scholars, scientists, and all people who develop a deep connection with their place on Earth.

Cutting of Marine Isthmuses

Besides the great enterprises of physical transformation of which I have already spoken, other works of internal improvement or change have been projected in ancient and modern times, the execution of which would produce considerable, and, in some cases, extremely important, revolutions in the face of the earth. Some of the schemes to which I refer are evidently chimerical; others are difficult, indeed, but cannot be said to be impracticable, though discouraged by the apprehension of disastrous consequences from the disturbance of existing natural or artificial arrangements; and there are still others, the accomplishment of which is ultimately certain, though for the present forbidden by economical considerations.

When we consider the number of narrow necks or isthmuses which separate gulfs and bays of the sea from each other, or from the main ocean, and take into account the time and cost, and risks of navigation which would be saved by executing channels to connect such waters, and thus avoiding the necessity of doubling long capes and promontories, or even continents, it seems strange that more of the enterprise and money which have been so lavishly expended in forming artificial rivers for internal navigation should not have been bestowed upon the construction of maritime canals. Many such have been projected in early and in recent ages, and some trifling cuts between marine waters have been actually made, but no work of this sort, possessing real geographical or even commercial importance, has yet been effected.

These enterprises are attended with difficulties and open to objections, which are not, at first sight, obvious. Nature guards well the chains by which she connects promontories with mainlands, and binds continents together. Isthmuses are usually composed of adamantine rock or of shifting sands—the latter being much the more refractory material to deal with. In all such works there is a necessity for deep excavation below low-water mark—always a matter of great difficulty; the dimensions of channels for sea-going ships must be much greater than those of canals of inland navigation; the height of the masts or smoke pipes of that class of vessels would often render bridging impossible, and thus a ship canal might obstruct a communication more important than that which it was intended to promote; the securing of the entrances of marine canals and the construction of ports at their termini would in general be difficult and expensive, and the harbors and the channel which connected them would be extremely liable to fill up by deposits washed in from sea and shore. Besides all this, there is, in many cases, an alarming uncertainty as to the effects of joining together waters which nature has put asunder. A new channel may deflect strong currents from safe courses, and thus occasion destructive erosion of shores otherwise secure, or promote the transportation of sand or slime to block up important harbors, or it may furnish a powerful enemy with dangerous facilities for hostile operations along the coast.

Nature sometimes mocks the cunning and the power of man by spontaneously performing, for his benefit, works which he shrinks from undertaking, and the execution of which by him she would resist with unconquerable obstinacy. A dangerous sand bank, that all the enginery of the world could not dredge out in a generation, may be carried off in a night by a strong river flood, or a current impelled by a violent wind from an unusual quarter, and a passage scarcely navigable by fishing boats may be thus converted into a commodious channel for the largest ship that floats upon the ocean. In the remarkable gulf of Liimfjord in Jutland, nature has given a singular example of a canal which she alternately opens as a marine strait, and, by shutting again, converts into a fresh-water lagoon. The Liimfjord was doubtless originally an open channel from the Atlantic to the Baltic between two islands, but the sand washed up by the sea blocked up the western entrance, and built a wall of dunes to close it more firmly. This natural dike, as we have seen, has been more than once broken through, and it is perhaps in the power of man, either permanently to maintain the barrier, or to remove it and keep a navigable channel constantly open. If the Liimfjord becomes an open strait, the washing of sea sand through it would perhaps block up some of the belts and small channels now important for the navigation of the Baltic, and the direct introduction of a tidal current might produce very perceptible effects on the hydrography of the Kattegat.

The Suez Canal

If the Suez Canal—the greatest and most truly cosmopolite physical improvement ever undertaken by man—shall prove successful, it will considerably affect the basins of the Mediterranean and of the Red Sea, though in
a different manner, and probably in a less degree than the diversion of the
current of the Nile from the one to the other—to which I shall presently
refer—would do. . . .

There is, then, no reason to expect any change of coast lines or of natural navigable channels as a direct consequence of the opening of the Suez
Canal, but it will, no doubt, produce very interesting revolutions in the animal and vegetable population of both basins. The Mediterranean, with
some local exceptions—such as the bays of Calabria, and the coast of Sicily
so picturesquely described by Quatrefages—is comparatively poor in marine vegetation, and in shell as well as in fin fish. The scarcity of fish in
some of its gulfs is proverbial, and you may scrutinize long stretches of
beach on its northern shores, after every wind for a whole winter, without
finding a dozen shells to reward your search. But no one who has not
looked down into tropical or subtropical seas can conceive the amazing
wealth of the Red Sea in organic life. Its bottom is carpeted or paved with
marine plants, with zoophytes and with shells, while its waters are teeming
with infinitely varied forms of moving life. Most of its vegetables and its
animals, no doubt, are confined by the laws of their organization to warmer
temperatures than that of the Mediterranean, but among them there must
be many, whose habitat is of a wider range, many whose powers of accommodation would enable them to acclimate themselves in a colder sea.

We may suppose the less numerous aquatic fauna and flora of the Mediterranean to be equally capable of climatic adaptation, and hence, when the
canal shall be opened, there will be an interchange of the organic population not already common to both seas. Destructive species, thus newly
introduced, may diminish the numbers of their proper prey in either basin,
and, on the other hand, the increased supply of appropriate food may
greatly multiply the abundance of others, and at the same time add important contributions to the aliment of man in the countries bordering on the
Mediterranean.

A collateral feature of this great project deserves notice as possessing no
inconsiderable geographical importance. I refer to the conduit or conduits
constructed from the Nile to the isthmus, primarily to supply fresh water to
the laborers on the great canal, and ultimately to serve as aqueducts for the
city of Suez, and for the irrigation and reclamation of a large extent of

desert soil. In the flourishing days of the Egyptian empire, the waters of the Nile were carried over important districts east of the river. In later ages, most of this territory relapsed into a desert, from the decay of the canals which once fertilized it. There is no difficulty in restoring the ancient channels, or in constructing new, and thus watering not only all the soil that the wisdom of the Pharaohs had improved, but much additional land. Hundreds of square miles of arid sand waste would thus be converted into fields of perennial verdure, and the geography of Lower Egypt would be thereby sensibly changed. If the canal succeeds, considerable towns will grow up at once at both ends of the channel, and at intermediate points, all depending on the maintenance of aqueducts from the Nile, both for water and for the irrigation of the neighboring fields which are to supply them with bread. Important interests will thus be created, which will secure the permanence of the hydraulic works and of the geographical changes produced by them, and Suez, or Port Said, or the city at Lake Timsah, may become the capital of the government which has been long so long established at Cairo.

Canal across the Isthmus of Darien

The most colossal project of canalization ever suggested, whether we consider the physical difficulties of its execution, the magnitude and importance of the waters proposed to be united, or the distance which would be saved in navigation, is that of a channel between the Gulf of Mexico and the Pacific, across the Isthmus of Darien. I do not now speak of a lock canal, by way of the Lake of Nicaragua or any other route—for such a work would not differ essentially from the canals, and would scarcely possess a geographical character—but of an open cut between the two seas. It has been by no means shown that the construction of such a channel is possible, and, if it were opened, it is highly probable that sand bars would accumulate at both entrances, so as to obstruct any powerful current through it. But if we suppose the work to be actually accomplished, there would be, in the first place, such a mixture of the animal and vegetable life of the two great oceans as I have stated to be likely to result from the opening of the Suez Canal between two much smaller basins. In the next place, if the channel were not obstructed by sand bars, it might sooner or later be greatly widened and deepened by the mechanical action of the current through it, and consequences, not inferior in magnitude to any physical revolution which has taken place since man appeared upon the earth, might result from it.

What those consequences would be is in a great degree matter of pure conjecture, and there is much room for the exercise of the imagination on the subject; but, as more than one geographer has suggested, there is one possible result which throws all other conceivable effects of such a work quite into the shade. I refer to changes in the course of the two great oceanic rivers, the Gulf Stream and the corresponding current on the Pacific side of the isthmus. The warm waters which the Gulf Stream transports to high latitudes and then spreads out, like an expanded hand, along the eastern shores of the Atlantic, give out, as they cool, heat enough to raise the mean temperature of Western Europe several degrees. In fact, the Gulf Stream is the principal cause of the superiority of the climate of Western Europe over those of Eastern America and Eastern Asia in the corresponding latitudes. All the meteorological conditions of the former region are in a great measure regulated by it, and hence it is the grandest and most beneficent of all purely geographical phenomena. We do not yet know enough of the laws which govern the movements of this mighty flood of warmth and life to be able to say whether its current would be perceptibly affected by the severance of the Isthmus of Darien; but as it enters and sweeps round the Gulf of Mexico, it is possible that the removal of the resistance of the land which forms the western shore of that sea, might allow the stream to maintain its original westward direction, and join itself to the tropical current of the Pacific.

The effect of such a change would be an immediate depression of the mean temperature of Western Europe to the level of that of Eastern America, and perhaps the climate of the former continent might become as excessive as that of the latter, or even a new "ice period" be occasioned by the withdrawal of so important a source of warmth from the northern zones. Hence would result the extinction of vast multitudes of land and sea plants and animals, and a total revolution in the domestic and rural economy of human life in all those countries from which the New World has received its civilized population. Other scarcely less startling consequences may be imagined as possible; but the whole speculation is too dreary, distant, and improbable to deserve to be long indulged in.

Covering Rock with Earth

If man has, in some cases, broken up rock to reach productive ground beneath, he has, in many other instances, covered bare ledges, and sometimes extensive surfaces of solid stone, with fruitful earth, brought from no inconsiderable distance. Not to speak of the Campo Santo at Pisa, filled, or at least coated, with earth from the Holy Land, for quite a different purpose, it is affirmed that the garden of the monastery of St. Catherine at Mt. Sinai is composed of Nile mud, transported on the backs of camels from the banks of that river. Parthey and older authors state that all the productive soil of the Island of Malta was brought over from Sicily. The accuracy of the information may be questioned in both cases, but similar practices, on a smaller scale, are matter of daily observation in many parts of Southern Europe. Much of the wine of the Moselle is derived from grapes grown on earth carried high up the cliffs on the shoulders of men. In China, too, rock has been artificially covered with earth to an extent which gives such operations a real geographical importance, and the accounts of the importation of earth at Malta, and the fertilization of the rocks on Mount Sinai with slime from the Nile, may be not wholly without foundation.

Wadies of Arabia Petræa

In the latter case, indeed, river sediment might be very useful as a manure, but it could hardly be needed as a soil; for the growth of vegetation in the wadies of the Sinaitic Peninsula shows that the disintegrated rock of its mountains requires only water to stimulate it to considerable productiveness. The wadies present, not unfrequently, narrow gorges, which might easily be closed, and thus accumulations of earth, and reservoirs of water to irrigate it, might be formed which would convert many a square mile of desert into flourishing date gardens and cornfields. Not far from Wadi Feiran, on the most direct route to Wadi Esh-Sheikh, is a very narrow pass called by Arabs El Bueb (El Bab) or, The Gate, which might be securely closed to a very considerable height, with little labor or expense. Above this pass is a wide and nearly level expanse, containing a hundred acres, perhaps much more. This is filled up to a certain regular level with deposits brought down by torrents before the Gate, or Bueb, was broken through, and they have now worn down a channel in the deposits to the bed of the wadi. If a dam were constructed at the pass, and reservoirs built to retain the winter rains, a great extent of valley might be rendered cultivable.

Incidental Effects of Human Action

I have more than once alluded to the collateral and unsought consequences of human action as being often more momentous than the direct and desired results. There are cases where such incidental, or, in popular speech, accidental, consequences, though of minor importance in themselves, serve to illustrate natural processes; others, where, by the magnitude and character of the material traces they leave behind them, they prove that man, in primary or in more advanced stages of social life, must have occupied particular districts for a longer period than has been supposed by popular chronology. "On the coast of Jutland," says Forchhammer, "wherever a bolt from a weak or any other fragment of iron is deposited in the beach sand, the particles are cemented together, and form a very solid mass around the iron. A remarkable formation of this sort was observed a few years ago in constructing the sea wall of the harbor of Elsineur. This stratum, which seldom exceeded a foot in thickness, rested upon common beach sand, and was found at various depths, less near the shore, greater at some distance from it. It was composed of pebbles and sand, and contained a great quantity of pins, and some coins of the reign of Christian IV, between the beginning and the middle of the seventeenth century. Here and there, a coating of metallic copper had been deposited by galvanic action, and the presence of completely oxydized metallic iron was often detected. An investigation undertaken by Councillor Reinhard and myself, at the instance of the Society of Science, made it in the highest degree probable that this formation owed its origin to the street sweepings of the town, which had been thrown upon the beach, and carried off and distributed by the waves over the bottom of the harbor." These and other familiar observations of the like sort show that a sandstone reef, of no inconsiderable magnitude, might originate from the stranding of a ship with a cargo of iron, or from throwing the waste of an establishment for working metals into running water which might carry it to the sea.

Parthey records a singular instance of unforeseen mischief from an interference with the arrangements of nature. A landowner of Malta possessed a rocky plateau sloping gradually toward the sea, and terminating in a precipice forty or fifty feet high, through natural openings in which the sea water flowed into a large cave under the rock. The proprietor attempted to establish salt works on the surface, and cut shallow pools in the rock for the evaporation of the water. In order to fill the salt pans more readily, he sank a well down to the cave beneath, through which he drew up water by a windlass and buckets. The speculation proved a failure, because the water filtered through the porous bottom of the pans, leaving little salt behind. But this was a small evil, compared with other destructive consequences that followed. When the sea was driven into the cave by violent west or northwest winds, it shot a *jet d'eau* through the well to the height of sixty feet, the spray of which was scattered far and wide over the neighboring gardens and blasted the crops. The well was now closed with stones, but the next winter's storms hurled them out again, and spread the salt spray over the grounds in the vicinity as before. Repeated attempts were made to stop the orifice, but at the time of Parthey's visit the sea had thrice burst through, and it was feared that the evil was without remedy. . . .

Every traveller in Italy is familiar with Monte Testaccio, the mountains of potsherds, at Rome; but this deposit, large as it is, shrinks into insignificance when compared with masses of similar origin in the neighborhood of older cities. The cast-away pottery of ancient towns in Magna Græcia composes strata of such extent and thickness that they have been dignified with the appellation of the ceramic formation. The Nile, as it slowly changes its bed, exposes in its banks masses of the same material, so vast that the population of the world during the whole historical period would seem to have chosen this valley as a general deposit for its broken vessels.

The fertility imparted to the banks of the Nile by the water and the slime of the inundations, is such that manures are little employed. Hence much domestic waste, which would elsewhere be employed to enrich the soil, is thrown out into vacant places near the town. Hills of rubbish are thus piled up which astonish the traveller almost as much as the solid pyramids themselves. The heaps of ashes and other household refuse collected on the borders and within the limits of Cairo were so large, that the removal of them by Ibrahim Pacha has been looked upon as one of the great works of the age.

The soil near cities, the street sweepings of which are spread upon the ground as manure, is perceptibly raised by them and by other effects of human industry, and in spite of all efforts to remove the waste, the level of the ground on which large towns stand is constantly elevated. The present streets of Rome are twenty feet above those of the ancient city. The Appian way between Rome and Albano, when cleared out a few years ago, was found buried four or five feet deep, and the fields along the road were elevated nearly or quite as much. The floors of many churches in Italy, not more than six or seven centuries old, are now three or four feet below the adjacent streets, though it is proved by excavations that they were built as many feet above them.

Nothing Small in Nature

It is a legal maxim that "the law concerneth not itself with trifles," *de minimus non curat lex;* but in the vocabulary of nature, little and great are terms of comparison only; she knows no trifles, and her laws are as inflexible in dealing with an atom as with a continent or a planet. The human operations mentioned in the last few paragraphs, therefore, do act in the ways ascribed to them, though our limited faculties are at present, perhaps forever, incapable of weighing their immediate, still more their ultimate consequences. But our inability to assign definite values to these causes of the disturbance of natural arrangements is not a reason for ignoring the existence of such causes in any general view of the relations between man and nature, and we are never justified in assuming a force to be insignificant because its measure is unknown, or even because no physical effect can now be traced to it as its origin. The collection of phenomena must precede the analysis of them, and every new fact, illustrative of the action and, reaction between humanity and the material world around it, is another step toward the determination of the great question, whether man is of nature or above her.

Suggested Readings

About George Perkins Marsh

Curtis, Jane, Will Curtis, and Frank Lieberman. 1982. *The World of George Perkins Marsh, America's First Conservationist and Environmentalist.* Woodstock, Vt.: The Countryman Press. A short and richly illustrated biography of Marsh that emphasizes the key events in his life.

Koopman, Harry L. 1892. *Bibliography of George Perkins Marsh.* Burlington, Vt.: The Free Press Association. A complete annotated list of all Marsh's publications.

Lowenthal, David. 1958. *George Perkins Marsh: Versatile Vermonter.* New York: Columbia University Press. An early scholarly biography of Marsh.

———. 2000. *George Perkins Marsh: Prophet of Conservation.* Seattle: University of Washington Press. Marsh's most recent biography, which interprets his life in light of conservation developments since Lowenthal's earlier treatment of Marsh.

About Environmental Change

Botkin, Daniel B. 1990. *Discordant Harmonies: A New Ecology for the Twenty-first Century.* New York: Oxford University Press. An exploration of different philosophies regarding the question, What is meant by the word "natural"?

Cronon, William. 1983. *Changes in the Land: Indians, Colonists, and the Ecology of New England.* New York: Hill and Wang. An environmental history that explores how Indians and European settlers transformed, and were transformed by, the New England landscape.

Judd, Richard W. 1997. *Common Lands, Common People: The Origins of Conservation in Northern New England.* Cambridge, Mass.: Harvard University Press. A detailed account of the history of conservation philosophies and natural resource management programs in New England.

Leopold, Aldo. 1949 [1966]. *A Sand County Almanac, with Essays on Conservation from Round River.* New York: Ballantine Books. A modern classic in conservation philosophy, which further develops the theme of humanity's relationship with the environment.

Marsh, George Perkins. 1864 [1965]. *Man and Nature: or, Physical Geography as Modified by Human Action.* (David Lowenthal, editor). Cambridge, Mass.: Belknap Press. The complete text of *Man and Nature,* with historical footnotes provided by Marsh's primary biographer.

Noss, Reed F., and Allen Y. Cooperrider. 1994. *Saving Nature's Legacy: Protecting and Restoring Biodiversity.* Washington, D.C.: Island Press. The most complete description available on the importance and design of ecological reserve systems.

228 *Suggested Readings*

Reisner, Marc. 1993. *Cadillac Desert: The American West and Its Disappearing Water* (2nd edition). New York: Penguin Books. An authoritative yet immensely readable account of the development of water control projects in the western United States. Reisner's story is essentially that of what happened because Marsh's advice about irrigation in the American arid West was ignored.

Whitney, Gordon G. 1994. *From Coastal Wilderness to Fruited Plain: A History of Environmental Change in Temperate North America from 1500 to the Present.* New York: Cambridge University Press. A general exploration of the ecological transitions that occurred in North America following its settlement by Europeans.

Williams, Michael. 1989. *Americans and Their Forests: A Historical Geography.* New York: Cambridge University Press. A detailed account of the history of the forest-products industry and forest conservation in the United States.